Trauma is intensified when reality is denied, minimised, or ignored. This radical and important book holds the reader's hand while looking at our peculiar, shared, global trauma squarely in the eye. With an Australian focus and a penchant for the truth, it carries for us the conversation that often fails to land with others.

Jean Renouf and Richard Hil deftly guide us along the path from anger to acceptance and the well spring of individual and collective agency, action and connection that sits beyond and alongside despair and grief as we 'creatively redesign' our future together.

For an exhausted community leader, this book has both fanned my activist flame and reminded me of the deep still pool that I can be for both myself and my community.

—Mel Bloor, *founder and president, Resilient Uki*

Brave, necessary, and beautifully human. This book is for you. *Unspeakable* is a courageous companion to a fierce subject: how we live with climate disruption without losing our humanity in a world that no longer feels like the one we once knew. It sits in the tension between hope and despair, honouring grief and anger while trusting that walking beside one another is not negotiable – but the only way through this transition. It invites you to move at the speed of trust, to act bravely without abandoning tenderness, and to remember that our shared courage grows stronger every time we show up for each other. You are seen here. And you are not alone. As Jane Fonda, climate activist and actor, said: 'Hope is a muscle, like the heart. It's when we take action that hope comes alive.'

—Wendy Brown, *President, Main Arm Disaster Recovery*

Climate change is the greatest threat facing humanity. Yet most political, business, finance and media leaders deliberately ignore its dire implications. Their insistence on pursuing unsustainable economic growth at all costs has now triggered the collapse of our previously stable climate, with widespread ramifications as safe planetary limits are exceeded.

The immediate impact shows up in the accelerating climate extremes, and related disasters, long evident globally. Leadership to address these catastrophes will not come from conventional institutions – it is coming from communities directly. This is why this book is long overdue. Jean Renouf and Richard Hil have direct hands-on experience of handling climate disasters, forming and motivating community groups to move beyond institutional inertia to positive action. But they cover much more than the immediate here-and-now, important though that is. They canvass the entire background to this existential threat, suggesting means of moving beyond feelings of impotence and rage, to see collapse as an opportunity to fundamentally change a flawed system into a sustainable and satisfying future – renewal within new limits. Thoroughly recommended!

—Ian Dunlop, *former Chair of the Australian Coal Association and CEO of the Australian Institute of Company Directors*

The effects of global warming are becoming more visible and alarming each year. This book is a field guide to life at the end of the world as we know it. It offers not false hope but sensible ways to cope in more resilient communities.

—Clive Hamilton, *Professor of Public Ethics, Charles Sturt University*

Unspeakable is an important contribution to the growing literature on living well together as the path through collapse to community. Weaving together threads of Indigenous wisdom, grassroots practice, spiritual thought, and academic rigour, Hil and Renouf tell a story that unflinchingly faces reality, ask crucial questions and provide vital guidance.

—Tim Hollo, author of *Living Democracy*

As we witness climate-fuelled worsening extreme weather driving deadly floods like those in Lismore in 2022, megafires like those in Maui and Los Angeles, and of course our own Black Summer, it would be easy to become paralysed with fear, and angered by the glacial pace of worldwide government efforts to reduce the belching of greenhouse gases into our already struggling, polluted atmosphere. Efforts ably delayed by entitled, self-absorbed but politically well-connected fossil fuel barons using their dirty money to drive a worldwide campaign of misinformation. Then along comes this book. While speaking honestly about what we all face, it provides hope rather than despair and throws us a lifeline – an alternative way of framing the existential threat of climate change, of reaching into the core of our humanity, and recreating true community. It tells that together we are much stronger. Richard and Jean's message, is essentially, 'It's OK, together we can do this'. Unspeakable is a must read.

—Greg Mullins *AO, AFSM. Former Commissioner Fire & Rescue NSW; founder of Emergency Leaders for Climate Action; author of* Firestorm, *and co-author,* The Fires Next Time

Unspeakable is one of those rare books that manages to bring together what is usually kept apart: science and emotion, spirituality and politics, inner life and agency, the individual and the collective, the dark and the light. Through a host of bold new concepts, this book enables us to fully engage with our times, with sensitivity, rigour, honesty and courage. One can feel the authors' deep affection for the world. Richard Hil and Jean Renouf guide us through the mourning of what is lost, without ever letting us sink into despair. For it is precisely there, in this difficult space, through connections and stories, through community, that something new and exciting is born. An essential read!

—Pablo Servigne, author of *Another End of the World is Possible: Living the Collapse (and Not Merely Surviving It)*

Whether we think in terms of collapse, the polycrisis, the great unravelling, or we just feel in our bones that all is not well, we know that we're collectively in a big mess. And we know that this mess is climate change and so much more. Our biggest threat as the unravelling builds may in fact be our normalcy bias – the false belief that things will revert to business as usual, despite the ever-shortening reprieve between local and global crises. It's time to make the unspeakable speakable – things really are as bad as we sense. Yet there is a way through, and we still have agency, choice and each other.

Unspeakable offers us the opportunity for honesty about our point in time – it brings the reality of this moment into clear perspective, while also offering guidance as to how we might move through and beyond it, together. And it's the latter that I most applaud, because knowledge without heart, and without signposts to help us navigate, will sink us. *Unspeakable* connects with our yearning to live differently, and it is this connection that can give us the juice we need to turn towards a future through and beyond what we now face – a future that could be regenerative, safe and resilient. It's not too late.

—Carmen Stewart *is a Regenerative Futures Practitioner who lives and works in the Northern Rivers of NSW, where she leads Imagine Northern Rivers*

Unspeakable

Richard Hil and Jean Renouf

Unspeakable

Facing up to the climate catastrophe and its consequences in Australia

PETER LANG

Oxford · Berlin · Bruxelles · Chennai · Lausanne · New York

Bibliographic information published by the Deutsche Nationalbibliothek
The German National Library lists this publication in the German National Bibliography; detailed bibliographic data is available on the Internet at http://dnb.dnb.de.

A catalogue record for this book is available from the British Library.

Library of Congress Cataloging-in-Publication Data
Names: Hil, Richard author | Renouf, Jean author
Title: Unspeakable: facing up to the climate catastrophe and its consequences in Australia / Richard Hil and Jean Renouf.
Description: Oxford; New York: Peter Lang, [2026] | Includes bibliographical references and index.
Identifiers: LCCN 2026017345 (print) | LCCN 2026017346 (ebook) | ISBN 9781803745442 paperback | ISBN 9781803745459 ebook | ISBN 9781803745466 epub
Subjects: LCSH: Climatic changes--Social aspects--Australia | Climate change adaptation--Australia | Human beings--Climatic factors--Australia | Australia--Environmental conditions--21st century
Classification: LCC QC903.2.A8 H55 2026 (print) | LCC QC903.2.A8 (ebook)
LC record available at https://lccn.loc.gov/2026017345
LC ebook record available at https://lccn.loc.gov/2026017346

Cover image: iStock credit: shaunl

ISBN 978-1-80374-544-2 (Print)
ISBN 978-1-80374-545-9 (ePDF)
ISBN 978-1-80374-546-6 (ePUB)
DOI 10.3726/b21994

Published by Peter Lang Ltd, Oxford, United Kingdom

info@peterlang.com

This publication has been peer reviewed.

www.peterlang.com

Contact for General Product Safety Regulation (GPSR): gpsr@peterlang.com

Contents

Acknowledgements

This book has been many years in the making. It draws on extensive primary and secondary research – interviews, books, articles, reports, and more – as well as countless conversations with colleagues, friends, and strangers over many years. We've also had the privilege of participating in campaigns, climate action and support groups, community education initiatives, and have established and ran a major community resilience organisation in the Northern Rivers region of NSW.

In all these spaces, we've rubbed shoulders with passionate, insightful, and committed individuals who, like us, are searching for answers to some very complex and urgent questions. Both of us have lived through extreme weather events and witnessed the pain and suffering they bring. Many people we know have endured such tragedies – often repeatedly. Their stories, their presence permeate every page of this book.

We are profoundly grateful to our families. To our wives, Jennifer, and Carly, for their unwavering support, integrity and steadiness. To our children, for whom this book – and all of our resilience work – is ultimately directed. They remain a daily reminder of what matters and why this work is necessary.

We also acknowledge Country, the custodians who have cared for these lands and waters since time immemorial; the more-than-human world that sustains us; and the future generations whose wellbeing guides so much of what we do.

Our gratitude extends to the many community members we have walked alongside, particularly the colleagues and friends at Plan C and Safer Future, whose commitment and care have shaped so much of what is documented here. We also thank our partners, funders, and donors who continue to believe in the value of community-led resilience. Deep appreciation is owed as well to community resilience leaders and supporters (you are too many to be listed here but you know who you are!), and to those emergency service personnel, as well as political, governmental, business and institutional actors who show clear-eyed leadership and a genuine willingness to pursue systems change.

And to the reader: thank you for your engagement, curiosity, and courage. You hold part of the response to what is happening around us, and your presence is woven into the life of this work. We recognise that some of what follows may be difficult to sit with, and we trust you to move at a pace that feels right – pausing when needed, stepping outside for air or perspective, and seeking the company of others where that helps. Attending to one's own wellbeing is part of this work. And if this book resonates with you, please share it with others.

Now, the usual rider applies: responsibility for what is written here rests solely with us – the authors. Still, we remain deeply grateful to the many people who have entered our lives and spoken candidly about the climate predicament and the prospect of societal collapse. Without their contributions, this book would never have come into being.

Special thanks go to Alexandre Carle, Andrew Russell, Jeremy Tager, Michael Shaw, Michelle Walter, Ken Golding, Stuart Rees, Gideon Polya, Richard Mochelle, Alex Pelizzon, Jennifer Grainger, Robyn Lincoln, members of an online collapse support group, and Ian Dunlop, Clive Hamilton, Lesley Hughes, and Mandy Nolan for their memorable observations and comments. We also extend our thanks to Dennis Sykes, Zjamal Xanitha, Paul Lambert, Erik Haan, Carmen Stewart, Peter Brennan, Chris Coffey, and Boyd Kellner for some immensely helpful conversations. Special thanks to Charlie Brennan for his left field interventions and making those around you think harder about praxis.

We also acknowledge the hundreds of conversations over the years with acquaintances at the Mullumbimby, New Brighton, and Murwillumbah markets – havens of robust and colourful exchange – and offer our thanks to the hundreds of workshop participants who shared their thoughts and feelings on climate-related matters.

Our thanks also extend to our interviewees, Cat Barker, Jane, Bobbi Allan, Gin Waters and Bec McNaught, and to our dear friend Jeremy Tager, who continues to rail against the injustices perpetrated by the powerful.

Finally, we offer our sincere appreciation to Laurel Plapp at Peter Lang for her patience, encouragement, and invaluable support throughout this project.

Introduction: Being in and of this moment

> The future is not what it used to be.
>
> —JÖRG FRIEDRICHS

> The most remarkable feature of this historical moment on Earth is not that we are on the way to destroying the world – we've actually been on the way for quite a while. It is that we are beginning to wake up, as from a millennia-long sleep, to a whole new relationship to our world, to ourselves and each other.
>
> —JOANNA MACY

Northern Rivers, 28 February 2022

On the morning of 28 February 2022, much of the Northern Rivers region of New South Wales – the place both authors of this book call home – woke to a landscape that no longer resembled itself. Overnight, entire towns and valleys had been swallowed by water and soil. Streets disappeared under currents that rose to heights few had imagined possible. Hillsides collapsed as a result of thousands of landslides and slips, reshaping the contours of the region in ways that remain visible today. Houses were observed floating away in the strong current. Cows were seen on house rooftops, carried there by surging waters. Small planes were found upside down. What had stood firm for decades gave way in a matter of hours.

The disruption was total. Roads were cut across the region, including the highway that carries most of the food, medicine, and supplies into the Northern Rivers. Supermarket shelves emptied within a day and stayed that way, in some places, for months. Power failed. Telecommunications were cut. Potable water ran short. Emergency services were immediately overwhelmed. Depots belonging to the State Emergency Services – the agency responsible for leading flood response – as well as fire and police stations were underwater. Hospitals and medical centres were inaccessible. Schools and workplaces closed, some indefinitely. Banks could barely operate; cash machines were dead. Some communities were cut off for days, others for weeks. For many, the outside world simply vanished.

The psychological shock of all this was immediate and profound – compounded by a second major flood only weeks later. Some residents lost their homes and/or livelihoods in the space of a single night; others lost neighbours, friends, or the landscapes that defined their sense of place. Yet even those untouched physically carried the imprint of the event – a collective jolt that rippled across the region. No one was spared the grief or disorientation that followed. The scale of loss, and the speed of it, unsettled our sense of safety and permanence in ways that continue to linger.

And yet, in the absence of functioning institutions, something extraordinary happened. In dozens of isolated pockets across the Northern Rivers, communities mobilised spontaneously. People who had never spoken to each other before, climbed into tinnies and kayaks, launched private boats, flew helicopters to evacuate people or drop goods, used tractors and machinery to reopen rural roads, improvised rope bridges, and went street by street to pull people from rooftops and out of attics. Rescue efforts unfolded from verandas, school ovals, community halls, car parks, makeshift helipads – anywhere dry ground could be found.

As the floodwater slowly receded, the response only grew. Donations poured in from across the region and beyond. Spontaneous community hubs formed, with people organising food distribution, cleaning crews, welfare checks, emergency accommodation, animal rescues, medication runs, and the clearing of unimaginable volumes of debris and mud. These hubs meshed with neighbourhood centres, community halls, and local councils, creating a decentralised yet remarkably coordinated response that carried many communities through the early months.

The trauma was deep and remains so. But the community response – its speed, its organisation, its sheer determination – was nothing less than astonishing.

Both authors of this book were involved in these efforts: as residents, as neighbours, as a firefighter, as people who led when needed and followed when someone else knew better. The experience was shattering, illuminating, and transformative. It revealed not only the fragility of the systems we rely on, but also the depth of collective care that remains possible, even amid devastation.

Our own response was shaped in part by prior experience. One of us, Jean Renouf, had previously worked internationally in disaster and conflict zones, and in 2019 founded Plan C ('our plan is the community'), a not-for-profit dedicated to strengthening community resilience, precisely because we anticipated that large-scale, cascading disasters – climatic, social, economic – would arrive in our region before our institutions were ready. We were acutely aware of the climate science, systemic risk, institutional fragility, and the urgent need to strengthen community resilience.

And still: the shock of 2022 got deep into our bones.

Despite our awareness, training, and experience, nothing buffered us from the trauma that followed – the exhaustion, the burnout, the grief for what was lost and what might still be lost. We witnessed firsthand the trauma etched on the faces of friends and acquaintances whose homes were caked in toxic mud, or who attended a free counselling service soon after the tragedy. We saw the tears, the anguish, the bewilderment, and shock – feelings viscerally familiar to us. We felt what thousands of others felt: the disorientation of realising that the systems meant to protect us could be washed away overnight. The realisation that – 'this is it', this is climate disruption in full view – a scenario of disaster that will reoccur many times, in worse ways, for the rest of our lives.

But we also witnessed something else: a communal spirit emerging from the chaos, a clearer picture of what must be done, what is possible, and what communities can build when they no longer have the luxury of waiting.

In the months and years that followed, we, alongside many others, worked tirelessly to support this emergence – through community organising, training, advocacy, research, and collaborations across the Northern Rivers and beyond.

This book sits within that wider arc. It is grounded in the climate science, in systems thinking, in the experiences of communities we have lived and worked in, and in our own vivid encounters with catastrophe. Our observations and insights flow from what we saw, what we learned, and what we are still learning.

Our hope is that the book offers readers a way to navigate the endings already underway – and the beginnings that might still be possible. What we lived through in the Northern Rivers was not an isolated event but a microcosm of something far bigger in existential and global consequence.

It's happening everywhere

Indeed, what unfolded in the Northern Rivers was not an anomaly, nor an aberration, nor merely a 'natural' disaster. It was one local expression of an accelerating global trend. As we make clear in Chapter 2, the climate catastrophe is a truly global phenomenon whose destructive effects are inescapable, though the severity of its impacts varies across regions and nation states.

The warnings about this unfolding catastrophe have been relentless for more than half a century. From *The Limits to Growth* in 1972 to *Project Drawdown* in 2017 to umpteen assessment reports from the Intergovernmental Panel on Climate Change (IPCC), and across a wider corpus of scientific, policy, and civil society alerts, the consequences of environmental degradation have been loud, clear, and remarkably consistent.

These warnings have been matched by an assortment of alternative economic and ecological models and perspectives which propose workable counters to economic growth, high productivity, extractivism, over-consumption and waste. These 'post-growth' orientations include the 'steady-state economy', 'de-growth', 'postgrowth / PostGDP', 'circular economy', 'doughnut economy', regenerative economies, ecoversities, ecovillages, ecological jurisprudence, transition towns and green new deal movements.[1] While many of these approaches share a commitment to living well within ecological limits, prioritising wellbeing over Gross Domestic Product, reducing material throughput, strengthening the commons, and designing economies that serve people and the planet rather the interests of the few, they often retain the vague hope that we can still escape environmental disaster and societal collapse. And yet, as systems analysts argue,[2] humanity now faces a stark bifurcation: either a Great Decoupling or a Great Collapse. Our view, based on our reading of various indicators, is that the latter scenario is the more likely outcome.

Throughout this book we refer interchangeably to the climate disaster, disruption, catastrophe, chaos, breakdown, and predicament, or simply to the 'Great Unravelling'[3] – terms which, for us, best capture the multifaceted nature of the ecological, social, political, and moral disintegration now shaping our lives. We are uncomfortable with words like 'crisis' and 'emergency', and especially 'change', because they appear to suggest the possibility of some sort of recovery and return to normalcy. These terms also centre response within the logic of control – implying that the same systems that created the catastrophe will be able to manage its consequences. We question such assumptions. There is no going back, no more

---

1 For an overview of post-growth scenarios see: Matthias Schmelzer, Andrea Vetter, and Aaron Vansintjan, *The future is degrowth: a guide to a world beyond capitalism* (London and New York: Verso Books, 2022). For a comprehensive overview of how ecological and environmental law reshape our understanding of the convergence of legal systems and ecological epistemologies, see Alessandro Pelizzon, *Ecological jurisprudence: the law of nature and the nature of law* (Singapore: Springer Singapore, 2025).

2 See: Wim Naudé, *Economic growth and societal collapse: beyond green growth and degrowth fairy tales* (Switzerland: Palgrave Macmillan, 2023).

3 The term 'the Great Unravelling' is widely used in relation to environmental degradation but is most well-known as a descriptor of economic and political decline in the US: Paul Krugman, *The Great Unravelling: losing our way in the new century* (New York: W. W. Norton & Company, 2003).

business as usual – despite relentless attempts to preserve the status quo.[4] We are entering – indeed we have already entered – the era of the post-Anthropocene where planetary changes are occurring beyond the limits of human control.[5] The Great Unravelling is not simply about the symptomatic effects of climate chaos; rather it reflects systemic breakdown in the form of ecological destabilisation, political and social fragmentation, economic volatility, declining institutional legitimacy, and the erosion of shared meaning. It is in every sense of the term, a defining moment in which familiar stories of progress, certainty, and continuity have lost their persuasive power.

Although this book begins in the Northern Rivers, we are acutely aware that countless others have had their worlds upended – in Australia and internationally. For many communities, particularly First Nations peoples, the experience of profound disruption did not begin with climate breakdown; it is part of a much longer story of dispossession, forced removal, cultural erasure, and ongoing structural violence. Yet Indigenous peoples have also shown extraordinary perseverance, resistance, and cultural continuity, carrying knowledge systems and practices that have sustained life through centuries of upheaval. Their survival is not only testament to resilience, but a reminder that the collapse of dominant systems does not affect all peoples in the same ways, nor begin at the same time. Acknowledging this history is essential for understanding the unequal burdens of the climate disaster, and for recognising the leadership and insight that Indigenous communities bring to imagining what comes next.

Facing mortality

Experienced most profoundly as a crisis of mortality – of bodies, species, cultures, ways of life and imagined futures – the Great Unravelling compels us to confront

---

4 World Wildlife Fund, 'UN report shows human activity has changed the planet forever', *World Wildlife Fund News Stories*, 29 August 2021, https://www.worldwildlife.org/news/stories/un-report-shows-human-activity-has-changed-the-planet-forever, accessed 12 February 2025; Canadian academic Naomi Klein's *This changes everything: Capitalism vs. the climate, 2014*, (London: Penguin Books) asserts that climate breakdown will permanently transform economies and societies and that concerted political action is needed to overturn and replace the system responsible for this situation.

5 This term signals the denial of human-centredness and affirms the autonomous actions of unleashed natural forces as well as the interconnected and reciprocal reality of living species. Anna Tsing examines how inter-species, regenerative life in the ruins of capitalism might force news ways of being around connection, interdependence, and cooperation. Anna Lowenhaupt Tsing, *The mushroom at the end of the world: on the possibility of life in capitalist ruins* (Princeton, NJ: Princeton University Press, 2015).

the fragilities that industrial modernity has long denied. Death of course, is nothing new, but the scale, pace, and injustice of environmental degradation, and the mortality it ushers, disrupts all our assumptions of life's journey. At the heart of many commonly held beliefs, particularly in rich western nations, is the promise that lifespans will expand, health will advance, and each new generation will inherit a richer, more stable foundation than the one before. This promise no longer necessarily holds.

Uncertainty and foreboding have increasingly become the twin pillars of the new zeitgeist. And as planetary destabilisation accelerates, so our temporal horizons retract. Climate projections for 2050, 2080, or 2100 no longer feel like distant abstractions but as grim thresholds within our own and our children's lifetimes. Our experience is akin to a patient confronted with a devastating diagnosis: shock, denial, bewilderment, and anger, alongside moments of clarity, revaluation, and even peace. Many of us oscillate between presence and avoidance – veering from grief and fear to denial and distraction, often in the same day.

The fact is that climate disruption has *already* escalated mortality, acute in some places, chronic in others. The precarity of the future – and the foreboding and anticipatory grief that goes with it – is increasingly part of our routinised experience. Extreme heat, storms, and floods have already caused widespread death and destruction, generating enormous loss and suffering, with the most vulnerable bearing the brunt. Australia's own 2019–2020 Black Summer bushfires made the scale of devastation clear: smoke and flames from the fires were linked to an estimated 417 excess deaths, thousands of hospitalisations, and billions of animal deaths.[6] In 2022, floods in Pakistan affected an estimated 33 million and killed around 1,760 people.[7] A year later, sudden floods in Libya killed around 5,000 people.[8] The death toll in Europe following severe heatwaves in 2003 was estimated at 70,000 people, and 60,000 in 2022 – this despite the implementation

6 Nerilie B. Arriagada, Adrienne J. Palmer, David M. J. S. Bowman, Grant G. Morgan, Bikie B. Jalaludin and Fay H. Johnston, 'Unprecedented smoke-related health burden associated with the 2019–20 bushfires in eastern Australia', *Medical Journal of Australia* 213/6 (2020), 282–283; WWF-Australia, '3 billion animals impacted by Australia's bushfire crisis', WWF-Australia (14 September 2020), <https://www.wwf.org.au/news/2020/3-billion-animals-impacted-by-australia-bushfire-crisis>, accessed 25 March 2025.

7 Editors of Encyclopaedia Britannica, 'Pakistan floods of 2022', Encyclopaedia Britannica (2024), <https://www.britannica.com/event/Pakistan-floods-of-2022>, accessed 25 March 2025.

8 United Nations Office for the Coordination of Humanitarian Affairs, Libya: Flood update Flash Update No. 2 (15 September 2023) (New York: UNOCHA, 2023) <https://www.unocha.org/publications/report/libya/libya-flood-update-flash-update-no2-15-september-2023>, accessed 25 March 2025.

of various heatwave action plans. Those most impacted by such events were the elderly, the poor, and people living in urban heat islands without access to cooling systems or healthcare.[9] Similarly, Hurricane Katrina in the US showed that Black and low-income communities were disproportionately affected, with higher mortality and displacement linked to racialised housing patterns, underinvestment in levees, and inequitable disaster response.[10] Tardy and inadequate responses in this case were structured around histories of racism, poverty, and political neglect. Cyclone Nargis in Myanmar (2008) killed around 140,000 people, not only as a result of the storm itself but due in large part to governmental suppression of warnings, lack of infrastructure, and delayed humanitarian aid.[11]

The fact that such carnage is barely acknowledged in more affluent nations like Australia reveals some troubling blind-spots in how we view the world. As Rob Nixon argues, this unequal exposure constitutes a form of 'slow violence': 'attritional violence that is typically not viewed as violence at all'.[12] Everyday exposures – excess heat, smoke, unsafe water, food insecurity – are illustrative of this concealment, and the abuses of power that are rarely reported or commented upon. Such harm is often obscured by more 'spectacular' disasters 'nearer to home' or buried under competing, relatively trivial news and infotainment. Deaths from climate disruption are frequently depicted as 'natural' events rather than the outcome of policy decisions, governmental neglect, legal yet harmful practices, and other systemic failures.[13]

These outcomes echo feminist philosopher Judith Butler's delineation of precariousness and grievability in which some lives are treated as less worthy of mourning than others. Similarly, political theorist Achille Mbembe's notion of 'necropolitics' illustrates how modern power structures create 'death worlds' through increasingly harsh border regimes and policing methods, as well as through underfunded services, and the general abandonment of 'disposable' populations.

Australia's first National Climate Risk Assessment, publicly released in late 2025, underscores these inequities, dramatic increases in heat-related deaths and

---

9 Daniel Hémon and Éric Jougla, 'Surveillance of heat-related mortality in France during the summer of 2003', Eurosurveillance 9/8 (2004), 6–8.

10 Chester Hartman and Gregory D. Squires, eds, *There Is No Such Thing as a Natural Disaster: Race, Class, and Hurricane Katrina* (New York: Routledge, 2006).

11 Mark Pelling and Kathleen Dill, 'Disaster politics: tipping points for change in the adaptation of sociopolitical regimes', Progress in Human Geography 34/1 (2010), 21–37.

12 Rob Nixon, *Slow Violence and the Environmentalism of the Poor* (Cambridge, MA: Harvard University Press, 2011), p. 2.

13 Nixon, Slow Violence and the Environmentalism of the Poor, 2.

cascading hazards, particularly among socially and economically marginalised communities.[14]

This underscores the point that climate breakdown is not only an ecological tragedy but a profound political and socio-economic injustice inflicted upon some of the most vulnerable populations.

Climate-related mortality is entangled with a wider systemic issue that threatens the conditions required for life to flourish. As ecosystems collapse, species disappear, soils degrade, and water systems destabilise, the biophysical foundations of human wellbeing are eroding at unprecedented speeds. These losses reverberate socially as communities splinter under the strain of repeated disasters and political fragmentation; people are displaced from places that hold memory and meaning; and stable futures become far harder to imagine. For Indigenous peoples in Australia and globally, the destabilising effects of climate change intersect with a much longer history of imposed mortality: frontier violence, displacement, disease, the destruction of food systems, and environmental degradation on Country. Climate disruption compounds these harms, undermining cultural continuity, damaging sacred places, and threatening life-sustaining ecologies.

Further, rising global tensions over resources, borders, and geopolitical influence filter down into local life, influencing how people and communities interact, who is protected, and who is left exposed. In this context, uncertainty itself becomes a kind of ambient hazard – fuelling anxieties and concerns over an increasingly bleak future.

On the 'unspeakability' of the climate catastrophe

Yet oddly – perversely perhaps – when it comes to the full, grim story of the climate catastrophe, a cultural silence tends to prevail, even in the face of the most alarming evidence. As Greta Thunberg so graphically put it in 2019, 'our house is on fire'.[15] In any other context, such a declaration would trigger an immediate mobilisation of firefighters, arriving within minutes. In assessing the situation, engineers would redesign buildings, governments would enforce new safety codes, and communities would demand action to prevent future occurrences.

---

14 Australian Climate Service (ACS), Australia's national climate risk assessment (Australian Government, 2025), available at <https://www.acs.gov.au/pages/national-climate-risk-assessment> (accessed 25 October 2025).

15 Greta Thunberg, 'Address at World Economic Forum: 'Our house is on fire' – Jan 25, 2019' (Archives of Women's Political Communication, 25 January 2019), <https://awpc.cattcenter.iastate.edu/2019/12/02/address-at-davos-our-house-is-on-fire-jan-25-2019>, accessed 26 October 2025.

Yet when it comes to climate breakdown – an unfolding disaster on a planetary scale – our collective response has been hesitant, inconsistent, or muted to the point of denial. For more than thirty years, nations have gathered at the United Nations' Conference of the Parties (COP), issuing repeated statements of concern while global emissions continue to rise, ecosystems collapse, and global heating intensifies. It is as if the magnitude of this situation renders us speechless, unable or unwilling to absorb the terrifying reality that is being presented to us.

The difficulty of speaking honestly about the climate disaster is compounded by the fact that many communities have already endured forms of collapse that went unacknowledged. Indigenous peoples in Australia and around the world have lived through profound ecological and cultural devastation for generations, often with little recognition from the societies that benefitted from it. Their histories reveal how easily loss can be normalised, embedded in policy, or rendered invisible – and how silence becomes a tool for maintaining harmful systems. Understanding these longer histories of erasure and endurance is essential for grasping why the climate catastrophe remains so hard to name, and why truth-telling is a necessary precondition for meaningful change.

The threat, perhaps, is too large, the implications too destabilising, the required transformations too vast for the political imagination. Inertia, self-delusion, delay and denial are the result. There are risks of course, in talking about this stuff. The increasing frequency of fires, floods, extinctions, displacements and ecosystems collapse, offer a terrifying glimpse of what lies before us. This unsettles the stories that have for so long shaped our lives – stories of progress, growth, security, and continuity. So, avoidance becomes a coping mechanism, and silence a form of collective and political self-protection. But the 'unspeakability' itself is part of the problem: what cannot be named cannot be confronted, and what is not confronted escalates. One of the main tasks we have set ourselves in this book is to make climate disruption and the potential for societal collapse more 'speakable' – to find a language that is honest, clear and grounded enough to help us navigate what is already here – and to come.

Yet, the future is not set in stone. To live in this era is to stand between a world that is unravelling and a world not yet born. The challenge we now face is not only to try and prevent further damage but to find ways to live meaningfully within what is emerging. As Professor Clive Hamilton rather colourfully observed during a Plan C event titled *Facing up: how do we live in the midst of the climate and ecological crises?*,[16] 'there is a difference between being fucked and being totally

16 Plan C, 'Facing Up: how do we live in the midst of the climate and ecological crises?', Plan C (no date), <https://www.planc.org.au/facingup>, accessed 26 October 2025.

fucked'. Provocative as it is, the remark captures the fragile space we now inhabit – a recognition that while the unravelling is happening, agency remains – we, the people have choices about how we will respond to what lies ahead.

This book recognises that the great challenges of our time – climate, biodiversity, inequality, housing, mental health, artificial intelligence – are the by-products of an industrial system rooted in colonialism and extractivism. The damage wrought by this system now compels us all to confront some pressing metaphysical and ontological questions: How do we live now? What does it mean to live a good life in a damaged world? How do we hold courage and grief together? The answers to these and many other questions will to a great extent determine the future paths we tread.

Our focus

Our primary concern in this book relates to the question of agency – not the individualised, self-help version that imagines solitary preppers, but the grounded, collective capacity to act in solidarity where we are. We ask what it means to exercise agency in the ruins of modernity: what we can do, together, here and now, without waiting for governments or institutions to lead. This is not a rejection of governmental responsibility – holding power to account remains vital – but an insistence that change also begins locally in the everyday. How can we live meaningfully and ethically in the midst of collapse? What forms of agency remain when faced with an overwhelming reckoning? Agency, in this sense is both practical and moral: it rests in how and where we deploy our energies, commitments and imaginations to shape what comes next. The choices and decisions we make can signal a different kind of power: one grounded in cooperation, care, in civic solidarity, and in the refusal to give up on the future. It is in this sense that the personal is indeed the political. We view agency in the context of the need to build a new politics of solidarity and reconnection in which rights and justice are expressed through mutual aid and shared understanding – understanding of our plight and a commitment to survive and perhaps thrive even in the darkest of times.

This book is not about 'solutions' in the conventional sense; it does not offer instructions, promise rescue or a return to normality. Instead, we propose ways of thinking, being, and acting that allow life to continue with dignity and purpose, leaving readers to decide how, and whether, these resonate within their own lives. Because the future is so uncertain and unknowable, and because correspondingly we must enter the world of the imaginary to make sense of what is before us.

Structure of the book

The book is divided into two parts. Part I, *The Age of Endings*, steps onto the terrain of systems that are unravelling. It asks how we arrived at this point and why, despite all the evidence to the contrary, we continue to live as if little or nothing has changed. Through personal stories, scientific testimony, and systemic critique, we uncover the layers of silence and denial that veil the true nature of our predicament. We approach the climate disaster not only as a biophysical unravelling, but as a rupture in meaning and human purpose. We contend, however, that a 'time of endings' does not simply signify a terminal void but rather the simultaneous arising of new and prefigurative ways of being that offer the possibility of meaningful life in an upended world.

Chapter 1, *Living in a time of endings*, develops the idea of the Great Unravelling and the unspeakability that surrounds it. Through Jeremy Tager's account and wider reflections on what constitutes societal collapse, we explore the tension between what we know and what we are able or willing to say out loud. We argue that climate disruption cannot be grasped within the same moral and intellectual frameworks that produced it, and instead we call for new ethical and relational orientations – including a rethinking of civic life, re-embracing the idea of the social, and building resilient communities capable of responding to what is to come.

Chapter 2, *The trauma, the science, the unravelling*, turns to those closest to the data: Earth and climate scientists. It illustrates how working routinely with environmental datasets has generated considerable personal trauma for those involved – mirroring the anguish felt by so many other people. The voices of scientists, however, are now being aired but only somewhat, heard. They speak of the gravity of our plight and the need for urgent action – calls that have often gone unheeded. We take a close look at what exactly the science is telling us, and why we remain on a trajectory toward chaos and collapse.

Chapter 3, *Business as usual? Hypernormalisation and the climate disaster*, takes up that question directly. Beginning with Cat Barker's experience of the floods in Lismore in 2022, it examines how 'business as usual' is sustained through processes of hypernormalisation: the collective performance of normality in a world that is coming apart. We map the intersections of power that bind governments, fossil fuel corporations, finance corruptions, universities, and the media, and show how subsidies, political donations, financial complicity, greenwashing, and manufactured consent uphold 'cannibal capitalism'. Against this backdrop, we challenge narratives that individualise responsibility by identifying the main drivers of the climate disaster – corporation, wealthy elites, and the institutions and values that enable them.

Part II, *Reimagining Life as We Fall*, turns towards the question of how, in the midst of upheaval, we can live with clarity, purpose and meaning. It begins from the premise that, while we cannot prevent all that is coming, we can choose how we respond. We recognise that there is an array of options to face the climate predicament, many of which must be led and implemented by governments and international and national institutions. But, in line with our aim to bring agency back to you, the reader, we have chosen to focus on what we, as individuals and members of our local communities, can do, with or without governmental and institutional leadership.

In Chapter 4, *Living with mortality: From despair to deep peace*, we illustrate how climate disruption unsettles our inner worlds, producing grief, anxiety, moral disorientation – but also moments of clarity, courage and conviction. Drawing on Bobbi Allan's perspective as an elder, we argue that radical acceptance of what is unfolding – rather than harbouring illusions or sinking into unresolved grief – can offer a way forward. In such tumultuous times, deep peace and inner stillness, presence rather than resignation, and practices of reflection, self-compassion, and grounded awareness can help individuals stay calm and steady. We approach this sort of emotional work not as methodologies of retreat or mere new age indulgence, but rather as a prerequisite for considered, ethical action.

Chapter 5, *Rebuilding civic culture: Resilience through community*, discusses the vital, life-giving importance of relationality in all its complex forms. Beginning with Gin Water's account of the 2019–2020 bushfires, we show how communities can organise and act faster and more effectively than formal institutions when disasters strike. Drawing on research and local examples from the Northern Rivers region of New South Wales – for instance, community resilience hubs, neighbourhood networks, First Nations custodianship, and umpteen networks and alliances – we illustrate how social cohesion, trust, and reciprocity are the building blocks of personal and community resilience. In seeking to avoid romantic ideas of community – the world trips too easily off the tongue – we argue for social investment in strong civic cultures, with the backing and support of state agencies and institutions, that together can offer protection and the possibility of renewal. Re-engagement in civic life – an important theme throughout this book – and the reaffirmation of the idea of society, we contend, will generate the sense of common purpose we will need in a destabilised world.

Chapter 6, *The way ahead: Exercising agency through collapse and renewal*, brings the book's arguments into practical focus. It reframes mitigation (greenhouse gas emissions reduction) as an ethical responsibility – not a guarantee of rescue, but as a means of reducing suffering. We also reinforce the urgent need for huge public investment in adaptation infrastructure, while also warning against technocratic and individualised approaches that deepen inequality. We thus also stress the importance of justice-centred, community-based approaches to preparedness, as well as ongoing commitment to place-based, regenerative practices that restore ecological and social balance – something that has been lost under the watch of industrial modernism.

Despite all our promptings for some kind of renewal, we remain honest about their limits. There can be no return to situation normal. Our argument for budding civic-minded, democratic and resilient communities is, above all, about strengthening our collective capacity to endure, care, and uphold dignity (and life) in a moment of unprecedented transition.

In the conclusion to this book, *Facing endings, cultivating beginnings*, we draw various strands together, reaffirming our central contention that we are at an inflection point which invites people and communities to rethink the possibilities of life in an upended word. If we cannot rely on old stories, we must create new ones capable of guiding us through an uncertain future. That story should include some practical tips about what each of us can do, today, to help build stronger and more resilient neighbourhoods and communities. We offer these not as prescriptions, but as ethical signposts pointing toward meaningful, action-based ways of co-creating resilient communities. While the future will be shaped by forces already beyond our control, there is nonetheless scope to reduce harm, strengthen civic ties, and cultivate regenerative possibilities. We may be in a time of endings, but even at the twelfth hour humankind may come to realise that renewal and inter-being are only possible when the system that brought us to this point is replaced by something much better.

PART I

The Age of Endings

We are living in an age when the stories that once held sway are breaking apart. Ideas of growth, progress and certainty no longer hold. A civilisation built on extraction and expansion, on idle promises of endless acquisition and technological mastery over everything, is consuming itself and driving us to the brink of planetary destruction. Old promises, certainties and claims ring increasingly hollow against the unfolding reality of collapsing ecosystems, political fragility and fraying social bonds. This is a time of endings: of old certainties, of illusions, of the myths and narratives that that for so long have sustained modern life.

Part I of this book enters this landscape of unravelling. It asks how we got here and why, and despite all evidence, why we continue to act as though little or nothing has changed. Through personal stories, scientific testimony, and systemic critique, the following three chapters trace the layers of silence, denial and obfuscation that cloak the enormity of our plight. They explore how the climate disaster is not only a biophysical catastrophe but a crisis of meaning – one that unsettles our deepest assumptions about who we are and our place in the world.

To confront collapse is to walk a deeply troubling path: between grief and clarity, fear and insight. Yet it is only by facing what is ending that we can begin to discern what might still be possible. In the pause for reflection between acknowledgement and renewal, we invite the reader to stand with us in this era of transformation – to see the ruins of the old order not as a final catastrophe, but as the threshold to a different way of being.

CHAPTER 1

Living in a time of endings

> Not everything that is faced can be changed, but nothing can be changed until it is faced.
>
> —JAMES BALDWIN

> Collapse is not the end of everything, but the end of a certain world. It opens the door to other possibilities.
>
> —PABLO SERVIGNE AND RAPHAËL STEVENS

Introduction

We are living in an age of endings. To assert such is to make the claim that a way of life is disappearing, to be replaced by another semblance of being which resides largely in the realms of the unknown. The word *endings* is also highly contentious – not only because it implies a kind of finality – possibly as the foundation for some form of renewal – but because it carries with it a host of cultural assumptions. It can, for instance, evoke a linear conception of time aligned with the disciplinary logics of industrial modernism, where time is seen as the inevitable march of moments, eras, and epochs that reinforce narratives of endless growth, progress, certainty, and order. This is the instrumentalised world of clocks, calendars, and regulatory oversight.

Such a conception sits in stark contrast to traditional cultures whose understandings of time are deeply embedded in the relationality of place, ancestral connection, and intergenerational dreamtime and other stories. Time in this sense is considered an *everywhen* – a relational orientation to Country held through kinship obligations, ecological cues, ceremony, and deep-time stories. This orientation, however, like that of temporal linearity, has been disrupted in the face of climate chaos and environmental degradation more generally. The rhythms of nature, the cyclical patterns of change that once felt predictable, no longer hold.

But that's not all. When we speak of *endings*, we must also consider what has already transpired over millennia. Many worlds have already been upended – and in some cases, erased – through colonialism and empire. History is littered with examples of this. And yet, despite attempts to extinguish Indigenous peoples through violence, dispossession, and the fiction of 'natural expiry', their persistence and resistance through the most horrific of times, is marked by ongoing presence and resurgence.

While acknowledging its obvious limitations, therefore, we nonetheless use the word *endings* to signify the gradual unravelling of the assumptions and modalities that have long sustained industrial modernism – its commitments to growth, accumulation, profit, and extraction. It also refers to the convergence of problems such as growing economic inequality, political polarisation, loss of faith in democratic institutions, growing authoritarianism and the splintering of social solidarities and connections – problems which are all exacerbated by galloping environmental destruction. So, for us, 'endings' reflects the irreversible rupture to a way of life – industrial modernism – *and* to the ecologies that have sustained life through the Holocene and beyond. Both of course, are deeply interrelated. In common with other writers, therefore, we acknowledge, and respond to, the 'long descent' into the 'coming dark age'.[1]

Across the world, societies are in the midst of this descent that signals not merely a moment of temporary disruption, but the unravelling of long-embedded systems. The acceleration of climate breakdown, biodiversity loss, and socio-political upheaval reveal that what is collapsing is not only the stability of the Earth's fragile systems, but the core ideas of industrial modernism – growth, accumulation, progress, technological supremacy, and certainty. The promise of perpetual expansion, mastery over nature and technological salvation now appears increasingly hollow against mounting evidence of planetary destruction.

Clearly, to speak of collapse is to enter difficult, even 'controversial' terrain. It demands that we resist both denial and false optimism, and engage openly and honestly with the implications of living in a time of upheaval. As we note in a review of the collapse literature, the notion of the Great Unravelling refers to more than environmental degradation: it signals the end of a certain way of life. Unpalatable and disturbing as this may feel – many of course, will welcome the end of a rapacious system – all the telltale signs of collapse are now plainly in view.

We begin however, with a simple but unsettling premise: namely, that the silence surrounding climate disruption is itself a symptom of a civilisation in

---

1 John Michael Greer, The Long Descent: A User's Guide to the End of the Industrial Age (Canada: New Society Publishers 2008) and Pelizzon, Ecological jurisprudence, p. 15.

throes of disintegration. Through the story of climate activist and writer Jeremy Tager – a self-described 'climate refugee' who has retreated to the rainforest highlands of northern Australia – we glimpse the emotional and existential dimensions of this predicament. His sense of exile, fear, and grief reflects a wider 'unspeakable knowing' – a growing awareness that ecological and social systems are collapsing, even though this remains strangely unspoken in civic life.

Drawing on Jeremy's account, we explore why, for many of us, climate disruption and the prospect of societal collapse are so difficult to contemplate. There is in this regard, a huge gulf between knowledge and experience that invokes a range of unhelpful responses, including paralysis, denial, and confusion, even as the evidence of collapse becomes undeniable.

From this, there emerges a broader argument: that the climate catastrophe cannot be addressed within the same ontological and epistemological frameworks that produced it. To speak meaningfully about climate breakdown requires not only new narratives, but repurposed ethical and relational orientations – ways of being that recognise the interdependence of life on Earth. Above all, we urge a radical rethinking of civic life and the building of resilient communities capable of responding to the coming storm – indeed, without a reinvigorated sense of social solidarity and connectedness we may face insurmountable odds.

This chapter therefore sets the stage for the book's most pressing inquiry: how to live, think, and act in a time of endings. We argue that acknowledging collapse does not foreclose the possibility of meaning or agency, but rather opens space for reimagining what it is to be human amid in a radically transformed world.

In the shadow world

Jeremy is the sort of person you want to spend time with – engaging, bright, and humorous. Yet, like the rest of us, he also has darker moments, especially when it comes to the state of the planet. A longtime follower of environmental science, Jeremy is acutely aware of what the evidence reveals. He describes himself as a 'climate refugee', now settled in a tiny hamlet on the Atherton Tablelands in Far North Queensland – one of the wettest places in Australia. His modest fibro home sits beside a patch of rainforest, which he tends with loving care. Occasionally, a baby cassowary emerges from the foliage to pay him a visit. Despite some early reservations, they've become good friends.

Jeremy's decision to move 1,800 kilometres from Uki, in the Northern Rivers region of New South Wales to the tropical north was driven by fear – particularly fear of fire.

> *I was more scared of the fires than the floods. I was living among the remnants of Gondwanan rainforests, and they're not meant to burn. While I was away in Western Australia for a year, fires swept through Uki [in 2019]. My home was at the end of a road – if fire came, there was nowhere to go. I'd have been toast. Although disruptive, flooding felt less of an existential threat.*

Jeremy recalls the deluge of 2022:

> *At least you can get to higher ground. You're not confronting a wildfire that moves faster than a car, with embers leaping kilometres ahead and igniting everything unpredictably. So, I started searching for a safer place. I was looking for wet places. Now, I live where it rains most of the year – four meters or more, right here on the Atherton Tablelands.*

But Jeremy knows his sense of safety is relative: '*Finding a safe home is only part of the problem. The planet isn't safe. The living world isn't safe. Friends, family, loved ones – none of us are safe*'. Witnessing the devastation of the Earth's ecosystems has led Jeremy down a path of deep sorrow:

> *It's about everything. First, the living world. It started with the Great Barrier Reef and a coral scientist who stated, matter-of-factly, that it was doomed. I fell into a depression deeper than any I'd ever known. Even here, in my "safe" home, creatures found nowhere else in the world are dying off. Some retreat to higher, cooler rainforest pockets, but those are mere slivers compared to what a thriving population needs. Trees are flowering at unpredictable times; migratory animals must recalibrate their journeys – if they can.*

Equally troubling is the deafening silence that shrouds our predicament – and the pervasive dissonance and denial. Our collective plight, Jeremy says, is unspeakable – buried somewhere deep in our unexpressed, grief-stricken being: '*In many cultures, grief is expressed publicly*' Jeremy continues,

> *But here? Even if there is grief, there's silence. Often, there isn't even that. Instead, there's denial, forgetting, ignorance, despair, anger turned inward – even baseless hope. And all of these serve as reasons not to acknowledge what's actually happening. This kind of silence isn't new – it's the same repression seen in memories of trauma or abuse, the refusal to recognize genocide in plain sight, the blind faith that convinces people to ignore reality. Sometimes it's privileged ignorance; sometimes, it's the fervour of those who believe they alone possess knowledge.*

For Jeremy, this tortuous silence – masked by numerous distractions – is obviously perplexing:

> *Climate disaster isn't something that will disappear if we close our eyes, shut the front door, or turn off the television. And yet, silence prevails. Sometimes, I feel like I'm living in a shadow*

world – one filled with people who exist in a reality where climate change isn't happening. It's the unmentionable house guest, too unsettling to acknowledge. Or perhaps, the terminal illness no one dares name, because the truth is too painful to bear.

An unspeakable knowing

At some point, however, this pain has to be faced, and talked about – however confronting that might be. Events will no doubt dictate how and when such conversations take place. It is no longer possible, we contend, to ignore the upending of industrial modernism, or indeed to deny the very real prospect of societal collapse. As Vanessa Machado de Oliveira remarks in her 2025 book, *Outgrowing Modernity: Navigating Complexity, Complicity, and Collapse with Accountability and Compassion:* 'The inevitable is coming fast. We know it in our bones – and its past time to face it'.[2] Confronted by this unpalatable vista, the late spiritual ecologist Joanna Macy argued that the greatest dangers of our times, 'is the deadening of our response'.[3]

Contemplating the destruction of Earth's fragile ecological systems, and therefore a way of life, is of course deeply confronting. It goes to the core of our being. How do we – indeed, can we – come to terms with this dark knowledge? Yet instead of squaring up to this terrifying prospect, many of us choose to cling to magical solutions, ill-conceived hope, and the elusive promise of return to some sort of normalcy. Those who accept the science – many still do not – struggle to process the irreversible changes that are reshaping the world.[4] The enormous complexity of our situation becomes, as Jeremy suggests, literally unspeakable – beyond our emotional capacities, language, and comprehension.

---

2 Vanessa Machado de Oliveira, *Outgrowing Modernity: Navigating Complexity, Complicity, and Collapse with Accountability and Compassion* (Berkeley: North Atlantic Books, 2025), 1.

3 Joanna Macy, 'The Greatest Danger', *YES! Magazine* (2 February 2008), <https://www.yesmagazine.org/issue/climate-solutions/opinion/2008/02/02/the-greatest-danger>, accessed 26 October 2025.

4 4 Caroline Fisher and Sora Park, 'The number of climate deniers in Australia is more than double the global average, new survey finds', *The Conversation* (16 June 2020), <https://theconversation.com/the-number-of-climate-deniers-in-australia-is-more-than-double-the-global-average-new-survey-finds-140450>, accessed 26 October 2025. The authors cite data in The Digital News Report: Australia 2020 indicating that climate denialism in Australia exceeds that of many other countries.

Sense-making and climate disruption

One major reason why climate disruption is so difficult to discuss is that it feels overwhelmingly complex. It is, after all, not a single event or issue, but a vast and complex network of interlinked temporal and spatial processes that confound ordinary human comprehension. Philosopher Timothy Morton refers to climate change as a 'hyperobject' – something so vast, persistent and widespread that it cannot be fully observed or understood from any single vantage point.[5] Like other phenomena such as relativity, dark matter and gravity, climate disruption stretches across various dimensions, existing everywhere and nowhere all at once, therefore eluding the ways in which humans usually make sense of the world around them. Cognitive linguist George Lakoff refers to this as 'hypocognition' – that is, the absence of conceptual frameworks to adequately describe or understand something so vast and amorphous.[6] This means that most people are lacking the mental and linguistic acuity to truly grasp what climate disruption actually means. The view is supported by psychology researchers Sabine Pahl and colleagues who point out that human cognition has evolved to prioritise immediate dangers – like predators or food shortages – rather than abstract, diffuse risks that unfold over decades. This evolutionary bias, they argue, makes it difficult to maintain focus or motivation around distant threats. Such dangers appear even more remote owing to the nature of modern life. More than half of humanity now reside in urban environments, where the rhythms of nature are muted or obscured. Disconnected from seasonal patterns and the intricacies of everyday 'natural' life, more and more people find it difficult, if not impossible, to notice subtle environmental changes or to link them to broader planetary shifts and changes.[7]

This mismatch between knowledge and experience is also captured in the conceptualisation of climate breakdown as a 'wicked problem', something that is considered so complex and dynamic that it eludes simple explanations or definitive solutions.[8] Wicked problems – such as climate breakdown – do not appear

---

5 Timothy Morton, *Hyperobjects: Philosophy and Ecology after the End of the World* (Minneapolis: University of Minnesota Press, 2013), 1.

6 George Lakoff, 'Why it Matters How We Frame the Environment', Environmental Communication: A Journal of Nature and Culture 4/1 (2010), 70–81.

7 Sabine Pahl, Stephen Sheppard, Christine Boomsma and Chris Groves, 'Perceptions of time in relation to climate change', *WIREs Climate Change* 5 (2014), 375–388.

8 Horst W. J. Rittel and Melvin M. Webber, 'Dilemmas in a general theory of planning', *Policy Sciences* 4/2 (1973), 155–169; Brian W. Head, 'Wicked Problems in Public Policy', *Public Policy* 3/2 (2008), 101–118; Catrien J. A. M. Termeer, Art Dewulf and Gerard E. Breeman, 'Governance of wicked climate adaptation problems', in Jürgen Knieling and Walter L.

to have clear boundaries or end points; any attempt to solve one aspect tends to create or reveal new dynamics elsewhere – a situation made more problematic as biospheric changes occur beyond the remit of human control. Some researchers have gone further, referring to climate disruption as a 'super wicked problem' because of its global complexity and resistance to simple solutions. It is simply too vast and complicated to comprehend, let alone to do something about it, fast.

To complicate matters even further, Lorraine Whitmarsh and Stuart Capstick observe, that for some people, a sense of relative comfort is derived from the view that climate disruption is a 'distant issue', something affecting faraway places or future generations rather than everyone in the here and now.[9] Its seemingly invisible and intangible nature – carbon in the atmosphere, melting permafrost, shifting ocean currents (even though such things can be measured) – makes it 'deeply un-relatable'. Even those directly impacted by extreme weather events struggle to link these to broader biospheric forces.[10]

The challenge of understanding seemingly impenetrable phenomena like climate disruption is compounded by other complexities – this time relating to various cultural and social influences. In another study conducted by Whitmarsh and Capstick, they note that people's beliefs and perceptions are shaped significantly through social interactions, particularly in the way family, friends, and communities discuss, rationalise and normalise the world around them.[11] Assumptions and understandings are frequently aired and reaffirmed in such social settings. The resulting internalisation of particular views and opinions become part of a person's identity, shaping how and what they present to the world. Knowledge acquired in the presence of significant others often carries a relational weight that supersedes other forms of understanding. As sociologist Andrew Hoffman points out, the thoughts and ideas integral to the formation of cultural identity often outweigh scientific understandings of climate disruption; that is to say, many people interpret information through the lens of their group affiliations

---

Leal Filho (eds), *Climate Change Governance* (Berlin: Springer, 2013), pp 27–39; Lorraine Whitmarsh and Stuart Capstick, 'Perceptions of climate change', in Susan Clayton and Christie Manning (eds), *Psychology and Climate Change: Human Perceptions, Impacts, and Responses* (Oxford: Academic Press, 2018), 13–33.

9 Whitmarsh and Capstick, 'Perceptions of climate change', 15.

10 Ezra M. Markowitz and Meaghan L. Guckian, 'Climate change communication: Challenges, insights, and opportunities', in Susan Clayton and Christie Manning (eds), *Psychology and Climate Change: Human Perceptions, Impacts, and Responses* (Oxford: Academic Press, 2018), 35–63.

11 Whitmarsh and Capstick, 'Perceptions of climate change', 15.

and personal and communal values.[12] Taking this a step further, behavioural scientists Ezra Markowitz and Meaghan Guckian demonstrate that individuals exhibit 'confirmation bias' when engaging with climate information – in other words, they often tend to accept evidence that aligns with their pre-existing worldviews and dismiss what challenges them.[13]

This may in part explain why extreme weather events are not always interpreted as signs of climate disruption. Instead, such occurrences are filtered through a variety of social influences, cultural norms and personal values:[14] one person may see a devastating flood as divine punishment, another as bad luck, and yet another as proof of global warming. In many societies – including Australia – debates about 'climate change' have often tended to be less about scientific assertions than competing moral, political, and cultural visions of the future in which biospheric changes can be subsumed by other concerns. That said, increasing evidence of climate upheaval can and often does cast doubt on previously held assumptions.

These sorts of socio-cultural influences mentioned above may also help explain why concern over climate breakdown varies so widely across countries and communities.[15] In some places, it is framed as an ethical or existential crisis; in others, it is treated primarily as an economic or ideological issue, or combinations of all these. Not surprisingly, this often generates polarisation and discord – divisions that are particularly evident in the United States, where belief in and responses to 'climate change' are sharply divided along partisan lines.[16]

One of the more usual default responses to the 'wicked problem' of climate disruption is to regard it as ultimately manageable and fixable. While it is considered complex, urgent, and resistant to simple solutions, with enough human ingenuity and financial backing, or so the argument goes, it can be addressed. This belief – commonly aired by governments, corporations and sections of the media – reflects an enduring faith in technological solutions, as well as a tendency to respond to immediate disasters rather than what is an evolving planetary predicament. Indeed, most of us tend to view climate breakdown as something that flares up now and again, only to recede into the background as more mundane and pressing things occupy our attention. Even when we do manage to set aside

---

12 Andrew J. Hoffman, *How Culture Shapes the Climate Change Debate* (Stanford, CA: Stanford University Press, 2015).
13 Markowitz and Guckian, 'Climate change communication', 38.
14 Whitmarsh and Capstick, 'Perceptions of climate change', 15.; Dan M. Kahan, 'Why we are poles apart on climate change', *Nature* 488 (2012), 255.
15 Whitmarsh and Capstick, 'Perceptions of climate change', 15.
16 Hoffman, *How Culture Shapes the Climate Change Debate*, 5.

time to consider such matters, our understanding of changing weather patterns is, as we have noted, contingent on a host of factors that can make climate breakdown feel less like a shared reality and more like a contested belief.

In the end, the lure of scientific facts can only go so far. For scientists and others seeking to cut through the fog, they must first navigate the complex terrain of culture, personal experience and ideology – as well as the not inconsiderable influence of politics and vested corporate interests.

That said, the increasing frequency and intensity of extreme weather events means it is getting harder to ignore climate breakdown, or to explain it away. We are becoming ever more aware that something fundamental has shifted – and shifted for good.[17] The world is not the same. It feels frighteningly uncertain – that at least, is what many of the polls and surveys seem to be telling us.[18]

The main challenge now in the climate action space, we contend, is to truly grasp the huge transformations that are reshaping life on Earth and to nurture adaptive responses that might offer protection, and the promise of renewal.

Despite the enormous challenges involved in this context, many of us remain stuck in various states of emotional and intellectual paralysis. We seem preoccupied with mundane concerns and are easily distracted by spectacle, consumption, and acquisition – or with simply getting by. As some observers noted of the Yellow Vest movement, which arose in response to France's 2018 proposal to raise fuel taxes in an effort to curb fossil fuel use, a tension of priorities exists between those concerned about the 'end of the world' and those worried about making it to the end of the month.

Perhaps not surprisingly, few of us fully grasp the magnitude of what we are facing and what this might mean in terms of everyday life. Yet the prospect of societal collapse – driven in large part by the climate disaster – is real and immanent. It cannot be ignored. As Zali Steggall, independent member for Warringah

---

17 Stefan Ellerbeck, 'Climate change is already altering everything, from fertility choices to insuring our homes', *World Economic Forum* (24 June 2022), <https://www.weforum.org/stories/2022/06/climate-change-weather-extreme-health>, accessed 27 October 2025. Ellerbeck reports on how people around the world are making decisions about their lives in the wake of a deteriorating climate.

18 The Australia Institute, 'All-time high of 75 % of Australians concerned about climate change', Media Release (13 October 2021), <https://australiainstitute.org.au/post/all-time-high-of-75-of-australians-concerned-about-climate-change>, accessed 27 October 2025. See also: Joëlle Gergis, Grant Blashki, Jane Gardner and Simon Bradshaw, *Climate trauma: the growing toll of climate change on the mental health of Australians* (Sydney: Climate Council of Australia, 2023), available at <https://www.climatecouncil.org.au/wp-content/uploads/2023/02/Report-Climate-Change-and-Mental-Health.pdf> (accessed 27 October 2025).

since the 2019 Australian federal election (when she defeated the incumbent, former Prime Minister Tony Abbott) notes, 'it's time to address how we actually survive climate change'.[19]

Amid the ruins of industrial modernity

In witnessing the end of the current order, and perhaps the emergence of new ways of being – the seeds of which were planted long ago in various 'alternative', 'unorthodox' and even revolutionary milieus – we face enormous, complex challenges. That at least, is the conclusion of many, like us, who see in the ailing system of industrial modernism all the telltale signs of a system in terminal decline. One horrifying symptom of this, of course, is the rapid emergence of a hothouse Earth – with everything this entails.[20]

For socio-cultural critic and co-author of *Uncivilization: The Dark Mountain Manifesto*,[21] Dougald Hine, climate breakdown tends to invite the usual technological, science-based solutionism that has long been part of the capitalist enterprise. Hine accepts that the climate catastrophe is radically reconfiguring life on Earth. However, in a later work, *At Work in the Ruins: Finding Our Place in the Time of Science, Climate Change, Pandemics and All the Other Emergencies*, he warns against the certainty of collapse, instead encouraging an acceptance – or 'surrender' – to the decline of industrial modernism.[22] Hine claims that he no longer dwells on the intricacies of the climate science, preferring to explore the emotions of loss and longing that it provokes, and to 'pause' before lurching into responses that repeat past errors.

While Hine rightly draws attention to the emotional complexities of bewilderment and loss – particularly in a grief phobic culture[23] – it remains absolutely

---

19 The New Daily, 'it's time to address how we actually survive climate change', *The New Daily* (9 November 2025), <https://www.thenewdaily.com.au/opinion/2025/11/09/cop30-australia-climate>, accessed 13 November 2025.

20 We take the term 'hothouse earth' from: Will Steffen, Johan Rockström, Katherine Richardson, Timothy M. Lenton, Carl Folke, Diana Liverman, Colin P. Summerhayes, Anthony D. Barnosky, Sarah E. Cornell, Michel Crucifix, Jonathan F. Donges, Ingo Fetzer, Steven J. Lade, Marten Scheffer, Ricarda Winkelmann and Hans Joachim Schellnhuber, 'Trajectories of the Earth System in the Anthropocene', *Proceedings of the National Academy of Sciences of the United States of America* 115/33 (2018), 8252–8259.

21 Paul Kingsnorth and Dougald Hine, *uncivilisation: the dark mountain manifesto* (online publication, 2009), <https://dark-mountain.net/about/manifesto>, accessed 28 October 2025.

22 Dougald Hine, *At work in the ruins: finding our place in the time of science, climate change, pandemics and all the other emergencies* (Vermont: Chelsea Green Publishing, 2023).

23 Weller, F. *The Wild Edge of Sorrow: The Wild Edge of Sorrow Official Workbook Reflections, rituals, and meditations for grief and renewal*, Berkley: North Atlantic.

essential in our view to continue to talk openly and honestly about the current and anticipated impacts of the climate disaster, if only to give communities a fighting chance of rebuilding fractured lives. The fact is that the climate is changing far more quickly than most scientists anticipated, meaning that the urgency of a coordinated response is greater than ever. As the 2020 Bushfire Royal Commission notes, 'some communities will have to cope with the effects of multiple nature hazard events at once, with the prospect of being affected by further hazard events before the recovery efforts have been completed'.[24] In other words, Australians will not just experience a disaster followed by several years to recover and prepare for the next one, but will increasingly have to face multiple, rolling and diverse disasters in short succession, making it increasingly hard to remain resilient.

In reflecting on our plight, Hine observes an 'ambiguity' in discussions about collapse. On the one hand they are concerned with humanity's survival (an anthropocentric view) while on the other, highlighting the end of a particular way of life.[25] In this book, we embrace this ambiguity when discussing existential threats. It's impossible to address the threat to life on Earth without considering the systems and practices that have led to this point. These are not mutually exclusive approaches; indeed, most deliberations on the threat posed by climate breakdown embrace this supposed tension.

We agree with Hine that our predicament needs to be understood through a historical lens and particularly the ideologies, systems and practices that have led us to this point. Over the course of three centuries or more, industrial capitalism has generated enormous wealth, albeit most of it expropriated from Indigenous peoples and unevenly distributed around the world.[26] Today's neoliberal iteration of this rapacious system has been built on the foundations of empire, colonial conquest, and the violent subjugation of First Nations peoples.

The so-called 'civilizing mission' that has legitimated this system is bound to a brutal, violent legacy of land theft, exploitation, ecological destruction, and attempted cultural erasure.[27] At its core, this system is predicated upon the

---

24 Royal Commission into National Natural Disaster Arrangements, *report* (Canberra: Australian government, 2020), 55, <https://www.royalcommission.gov.au/system/files/2020-12/RoyalCommissionintNationalNaturalDisasterArrangements-Reportaccessible.pdf>

25 Hine, *At work in the ruins*, 127.

26 One of the most celebrated analyses of the consequences of capitalism is Thomas Piketty, *Capital in the twenty-first century* (Cambridge, MA: Harvard University Press, 2014).

27 The term 'civilising process' was popularised by Norbert Elias in a 1939 book of the same term, *The civilizing process (Basel: Haus zum Falken, 1939).* It has since been used widely in the social sciences and other bodies of literature, often as a way of describing the material, religious and racialised justifications for colonial domination.

idea of dominion over nature, people and place.[28] It sanctifies private property, celebrates extractive agriculture, and promotes endless consumption, limitless growth, and capital accumulation. Overshoot and environmental degradation are its grim legacies. It is a system propped up by a global alliance of corporate power, financial elites, and complicit governments who continue to peddle the myth that 'there is no alternative' – a questionable proposition referred to by British social philosopher, Mark Fisher, as 'capitalist realism'.[29]

Although we are repeatedly told that capitalism has achieved great progress – more wealth, less poverty, longer lives[30] – these gains have come at a staggering cost: rising inequality, the hollowing out of communities and public services, and the relentless assault on the environment.

The climate catastrophe is one more horrifying symptom of this rapacious way of life. And yet, as Hine notes, the historical roots of our collective predicament are largely unfamiliar to people in 'advanced' nations like Australia. Few can trace the conditions of everyday life to the violent foundations of the capitalist nation-state, or trace today's environmental problems to a system that appears hell-bent on self-destruction.

What triggers societal collapse?

The spectre of societal disintegration has long haunted the collective imagination, as evidenced in films like *The Road*, *Don't Look Up*, *Sapiens*, *Carbon Dating*, and *Civil War*, or television series like *The Handmaid's Tale* and *The Last of Us*. Literary works including Tim Winton's *Juice*, Octavia Butler's *Parable of the Sower*, *The Overstory* by Richard Powers, *The Swan Book* by Alexis Wright, *Clade* by James Bradley, and Paolo Bacigalupi's *The Water Knife* and numerous plays and visual art forms depict dystopian scenarios where collapse is not merely plausible, but

---

28 See Alyssa Battistoni, *Free gifts: capitalism and the politics of nature* (Princeton, NJ: Princeton University Press, 2025).

29 Mark Fisher, *Capitalist realism: is there no alternative?* (London: Zero Books, 2009).

30 Some prefer to refer to the epoch of modernism rather than capitalism per se, arguing that enlightenment values of reason, science and humanism have ushered great human progress. See: Steven Pinker, *Enlightenment now: the case for reason, science, humanism, and progress* (New York: Viking Press, 2018). For critique of Pinker's thesis, see: Jeremy Lent, 'Steven Pinker's ideas are fatally flawed. These eight graphs show why', *Open Democracy* (21 May 2018), <https://www.opendemocracy.net/en/transformation/steven-pinker-s-ideas-are-fatally-flawed-these-eight-graphs-show-why/>, accessed 12 March 2025.

inescapable.[31] For those cognizant of what the climate science is reporting, such accounts are especially chilling – portents of what is already here, and to come.

Yet there is nothing new about societal collapse – there are countless examples of such over the course of human history. Indeed, the disintegration of civilisations is a recurring phenomenon, well-documented in both academic research and historical records. One major conclusion, however, emerges from this body of work; namely, that societal collapse rarely manifests as a sudden, cataclysmic event. Rather, it is a *process* – sometimes abrupt, sometimes gradual – inevitably marked by great upheaval and considerable human suffering. Civilisations seldom disappear overnight. More often, they slowly unravel, with moments of acceleration. In the early stages, we might have little idea that the world around us is falling apart.

In an attempt to map the origins and nature of societal collapse, Associate Professor at the University of Skövde, Sweden, Danilo Brozović, has conducted a comprehensive review of the collapse literature, encompassing 361 academic articles and 73 books.[32] His findings highlight five primary strands of inquiry: the study of historical collapses; theoretical explanations for societal failure; explorations of possible alternatives or pathways to avoid collapse; fictional representations of disaster; and projections concerning future climate-driven scenarios. These areas of inquiry have contributed greatly to our understanding of the antecedents of civilisational decline. Brozović focusses on numerous accounts of societal disintegration, ranging from Edward Gibbon's 1776 exploration of Rome's fall – attributed to moral decadence and barbarian incursions – to Thomas Malthus's warnings in 1798 regarding overpopulation, as well as more contemporary studies like Joseph Tainter's application of complexity theory and Jared Diamond's ecological explanations.

Each body of work, Brozović argues, holds valuable clues as to the complex, multifaceted nature of collapse. Invariably, he contends, societal decline is the result of a confluence of internal fragilities (concentration of elite power, corruption, deepening social, cultural divides, environmental hazards, etc.) and external pressures (war, invasion, mass migration, pandemics, economic shocks, market collapse).

Like Brozović, Research Fellow at the Centre for the Study of Existential Risk at the University of Cambridge, Luke Kemp, has conducted his own 'historical

---

31 Such films and books are part of a vast body of work in the creative arts which ponders the realities of dystopian futures, including those caused by environmental destruction.

32 Danijel Brozović, 'Societal collapse: a literature review', *Futures* 145 (2023), article 103075.

autopsy', comparing past and present forms of societal collapse.[33] He confirms many of the causes identified by Brozović but sees in today's situation a much graver *globalized* process of disintegration. In a 2019 essay for BBC online, Kemp defines collapse as, 'a rapid and enduring loss of population, identity, and socio-economic complexity. Public services crumble and disorder ensues as government loses control of its monopoly on violence'. According to Kemp, one of the key drivers of this scenario is the concentration of elite power which, owing largely to the extraction of 'more wealth from the people and the land', renders societies, 'more fragile, leading to infighting, corruption, immiseration of the masses, less healthy people, overexpansion, environmental degradation and poor decision making by a small oligarchy'.[34]

Examples of such oligarchal power are evident around the world, today. Indeed, as more and more nations lurch toward authoritarianism and therefore a greater concentration of elite power and wealth, societies are made less resilient and responsive, and therefore, are more vulnerable to shocks and upheavals. One of the most significant sources of destabilisation, says Kemp, is the disruption to supply chains brought on by extreme climatic disruptions and related environmental and economic causes.

The interdependent nature of the global economy makes social instability and chaos more likely. A 'tipping point' is reached, Kemp argues, 'when compounding stressors overrun societal coping capacity'.[35] For Kemp, the signs of collapse are everywhere: '... the world is worsening in areas that have contributed to the collapse of previous societies. The climate is changing, the gap between the rich and poor is widening, the world is becoming increasingly complex, and our demands on the environment are outstripping planetary carrying capacity'.[36]

For Kemp then, climate disruption is one significant element in a constellation of forces that are propelling the current system toward collapse. Environmental degradation – most visibly the accelerating impacts of climate disruption – has become a powerful catalyst deepening the political, economic, and social fractures already embedded within our institutions. The most pressing challenges we face – rising economic inequality, democratic erosion, intensifying political

---

33 Luke Kemp, *Goliath's curse: the history and future of societal collapse* (New York: Viking, 2025).

34 Luke Kemp, 'Are we on the road to civilisation collapse?', *BBC Future* (19 January 2019), <https://www.bbc.com/future/article/20190218-are-we-on-the-road-to-civilisation-collapse>, accessed 12 March 2025.

35 Kemp, 'Are we on the road to civilisation collapse?'.

36 Kemp, 'Are we on the road to civilisation collapse?'.

polarisation, the resurgence of authoritarian and securitised responses, social atomisation, and the risks posed by emerging technologies – are all poised, care of climate breakdown, to worsen, driving our systems toward eventual breakdown.

In tracking how climate disruption is impacting Australia, we can see its ripple effects right across the system. Water shortages, owing to drought and other factors, for example, have already been evidenced in repeated crises in the Murray–Darling Basin, ongoing groundwater depletion, and the reintroduction of water restrictions in multiple states. Climate disruption also leads to crop failures and increasingly unstable food yields, particularly affecting staple crops such as wheat, rice, and maize, with Australian wheat belts experiencing greater yield variability due to heatwaves and prolonged drought. At the same time, ecosystem loss – including coral reef die-off, collapsing fisheries, and widespread species decline – undermines food security and rural livelihoods, most visibly through repeated mass bleaching of the Great Barrier Reef and declining fish stocks along Australia's coasts. These pressures are compounded by soil degradation, as erosion, salinity, and variable climate make farming progressively more difficult in drought-prone agricultural regions.

Together, these trends intensify food and resource stress, resulting in rising prices, increased volatility, sudden price spikes, and periodic shortages. In Australia, this is reflected in sharp increases in food prices following floods and droughts that disrupt supply chains. A warming climate also places increasing strain on energy systems, with instability appearing in the form of blackouts, fuel shortages, and energy price shocks – illustrated in Australia by heatwave-driven spikes in electricity demand and grid stress during extreme summer conditions.

At the same time, various levels of government in Australia face increasingly severe economic strain, requiring substantial budgetary readjustments as the costs of climate-related infrastructure damage and long-term adaptation mount. In the wake of repeated climate disasters in recent years, government coffers have come under immense pressure from multi-billion-dollar public expenditures following catastrophic bushfires and floods. Repeated often enough, such calamities will have a profound impact on the ability of governments to meet many of their existing obligations, feeling likely increased public unrest and political hostilities. Additional pressure on government finances arises from increased migration and displacement, as populations are forced to move due to heat, flooding, drought, or the loss of livelihoods, including internal displacement from flood-affected communities and fire-ravaged towns. These dynamics place intense pressure on public services such as healthcare, housing, education, road and transportation networks, and emergency response systems, while also undermining social cohesion.

Climate destruction further exacerbates existing inequalities, with costs falling disproportionately on poor and marginalised communities. In Australia, this burden intensifies as insurers retreat, withdrawing coverage or dramatically raising premiums in bushfire- and flood-prone regions, leaving households, farmers, and local economies increasingly exposed to repeated losses.

The most conspicuous signs of societal breakdown, however, arise first in governance, as austerity measures bite and security and control are strengthened. Extremist movements gain traction, social tensions rise, and migrants or refugees are scapegoated for deeper structural failures. In response, governments increasingly turn to authoritarian measures, with emergency powers introduced during climate-related crises – such as disaster declarations and expanded policing – at risk of becoming normalised and permanent features of governance.

Many of these climate-driven developments have occurred in Australia. Given the baked-in alterations to biospheric conditions – as noted in Chapter 2 – all the effects itemised above will drive societies to the edge of collapse. No one, not even Kemp, is totally clear about what the realities of collapse will look like. There are too many imponderables here to draw any firm conclusions. The dissolving of centralised authority, albeit under more stringent forms of coercion and control, along with the fragmentation of communities – in some cases, disappearing altogether – will likely give rise to chaos, disorder as well as the search for ways of binding social conventions in order to establish liveable zones of habitation. Despite the scale of thew threats before us, Kemp argues in his 2025 book, *Goliath's Curse: The History and Future of Societal Collapse,* that political will, social resilience, new technologies and more social justice might still prevent the worst from occurring[37] – a proposition questioned throughout this book. The system, we argue, is already in its death throes. As post-humanist thinker and essayist Bayo Akomolafe asserts, 'The greatest challenge we face is a philosophical one: understanding that this civilization is already dead'.[38]

In an equally assertive account of the seeming inevitability of collapse in *How Everything Can Collapse*, French 'collapsologists' Pablo Servigne and Raphaël Stevens identify three key drivers contributing to the decline of industrial capitalism: 1. the depletion of fossil fuels leading to economic stagnation; 2. ecological disruptions such as mass migrations and biodiversity loss which exacerbate

37 Kemp, *Goliath's curse*, Chapter 7.

38 Bayo Akomolafe, *What climate collapse asks of us* (Bayo Akomolafe website, 2019), <https://www.bayoakomolafe.net/post/what-climate-collapse-asks-of-us>, accessed 14 March 2025.

conflict and disease; 3. and, the fragility of interdependent infrastructures – particularly those related to food, water, and energy – that risk cascading failures when disrupted.[39] Like Kemp, Servigne and Stevens argue that these dynamics are unfolding within a highly interconnected global system, making collapse more widespread than in previous times.

These structural problems, we contend, are in many ways compounded – particularly in western societies – by hyper-individualism, growing social disconnection and loneliness, as well as anxieties relating to social media, artificial intelligence and transhumanism – all of which place further strain on an already threadbare social fabric.[40] The upshot is that populations in such societies – bereft of much-needed support systems – are rendered more vulnerable to the challenges that the climate catastrophe will present.

Collapse, then, is not a distant threat – it is an immanent reality, manifesting in uneven and fragmented ways across the world. As the late Earth system scientist Will Steffen stated in 2020, 'given the momentum in both the Earth and human systems, and the growing difference between the "reaction time" needed to steer humanity towards a more sustainable future, and the "intervention time" left to avert a range of catastrophes in both the physical climate system (e.g., melting of Arctic sea ice) and the biosphere (e.g., loss of the Great Barrier Reef), we are already deep into the trajectory towards collapse'.[41] Professor Hans Joachim Schellnhuber, founder and former director of the Potsdam Institute for Climate Impact Research, has cautioned that exceeding a 2°C rise in global temperatures could set off a cascade of tipping points and feedback loops, potentially propelling the planet towards 4°C of warming. Such a scenario, he warns, could mark the collapse of human civilisation as we know it.[42] Tipping points occur when a threshold is reached, leading to rapid and irreversible changes that do not respond to reductions in greenhouse gases. This can trigger feedback loops, for example

39 Servigne and Stevens, *How everything can collapse*.

40 See Tomas E. Ström, 'Digital disenchantment: Postcapitalism, AI and the end of the world', *Arena Quarterly* 5 (2021), 35–42. The annual HILDA survey (Household, Income and Labour Dynamics in Australia Survey in 2025 showed a marked increase in loneliness especially among younger men, much of this being put down to cost-of-living constraints and over-reliance on social media. See Iain Caldwell, 'Loneliness epidemic in Australia: Australians finding it harder to make new friends', *The Guardian* (20 September 2025), <https://www.theguardian.com/society/2025/sep/20/loneliness-epidemic-australians-finding-it-harder-to-make-new-friends-hilda-survey>, accessed 15 October 2025.

41 Asher Moses, 'Collapse of civilisation is the most likely outcome: top climate scientists', *Resilience* (8 June 2020), <https://www.resilience.org/stories/2020-06-08/collapse-of-civilisation-is-the-most-likely-outcome-top-climate-scientists/>, accessed 15 October 2025.

42 Moses, 'Collapse of civilisation is the most likely outcome'.

where melting ice reduces reflectivity, causing more heat absorption and further melting. Recent evidence also suggests that warm-water coral reefs are the first major Earth system tipping point to have been crossed, impacting millions of people who rely on reefs for food, livelihoods, and coastal protection. Parts of the Greenland and West Antarctic ice sheets may already have reached tipping points which will lead to metres of sea-level rise. The global temperature, is breaching the 1.5°C limit set by the Paris Agreement, with the result of increasing the risk of triggering multiple tipping points, including Amazon rainforest dieback and collapse of the Atlantic Meridional Overturning Circulation, leading to the further destabilisation of ecosystems, food systems, and climate regulation.[43]

Echoing this dire outlook, Johan Rockström, who leads one of Europe's foremost climate research institutions, noted in 2019 that a world warmed by 4°C would likely be unrecognisable and deeply unstable. It might struggle, he suggests, to sustain even a billion people, with only a privileged few maintaining modern living conditions amid widespread upheaval and conflict.[44]

Scientific models indicate, as we note in Chapter 2, that what lies ahead – what has in fact already started – is an extended period of both biospheric and social disruption and planetary transformation. Ecosystems collapse, a gradual erosion of centralised authority and effective governance, and fragmentation of communities, resulting in social breakdown, discord and potentially, civil unrest and state-sanctioned violence, are the most likely outcomes of this disintegrative process.[45]

---

[43] See: ScienceDaily, 'Earth has hit its first climate tipping point, scientists warn', 29 October 2025, available at <https://www.sciencedaily.com/releases/2025/10/251029002920.htm> (accessed 30 November 2025); Carolyn Gramling, 'Coral collapse signals Earth's first climate tipping point', *Science News*, Washington DC, 12 October 2025, available at <https://www.sciencenews.org/article/coral-collapse-climate-tipping-point> (accessed 30 November 2025); Earth.Org, 'New climate report warns of imminent climate tipping point as emissions soar', Hong Kong, 18 June 2025, available at <https://earth.org/new-climate-report-warns-of-imminent-climate-tipping-point-as-emissions-soar/> (accessed 30 November 2025); Climate Analytics, *Rescuing 1.5°c: new evidence on the highest possible ambition to deliver the Paris Agreement*, Berlin, 6 November 2025, available at <https://climateanalytics.org/publications/rescuing-1-5c> (accessed 30 November 2025).

[44] Moses, 'Collapse of civilisation is the most likely outcome'.

[45] See: Tim Di Muzio, *Carbon capitalism: energy, social reproduction and world order* (London and New York: Rowman & Littlefield International, 2015); Jeff Goodell, *The heat will kill you first: life and death on a scorched planet* (New York: Little, Brown and Company, 2023).

Facing the unthinkable

The accumulated evidence now suggests that collapse is not a speculative horizon but an unfolding condition of the present. As the ecological, political, and social fault lines of industrial modernity widen, the question is no longer *if* collapse will occur, but *how* we will live in the midst of it. Scholars such as Pablo Servigne and Raphaël Stevens, central figures in the field of collapsology,[46] describe themselves not as prophets of doom but as 'realists' willing to confront this truth without evasion. For them, 'to be catastrophist' simply means to resist denial, to see clearly the disintegration already underway and to accept what will inevitably be lost.[47] Acceptance, in their view, is not resignation but the precondition for moral clarity, courage, and collective preparation.

This shift – from asking how collapse might be prevented to how it might be lived through – marks a profound turning point in public and scholarly discourse. It calls for a reframing of ethics, politics, and even ontology: what forms of life, relationship, and responsibility remain possible when the project of endless growth progress have run their course? As the protective myths of modernity unravel, more people are beginning to confront these questions, often haltingly, sometimes in silence. The task, as this chapter has suggested, is not to deny the unthinkable but to learn how to reflect and act within its daunting parameters.

Thus far, such huge expectantly questions remain unaddressed, mainly because we have yet to acknowledge the possibility of an end to the world as we – the offspring of modernism – have come to know it. To do so, certainly in main-stream political discourse, would be to admit failure and defeat. The dominant assumption in Australia, as noted, is that we can still get out of this almighty mess.[48] Few if any political leaders, however, talk openly about the irreversibility of biospheric changes let alone the possibility of societal collapse – to do so would be career-ending. The exceptions to the hegemonic line – found in documentaries, investigative reports, warnings from experts – stand in stark contrast to what is, in effect, a broad cultural silence about the gravity of our situation.[49]

---

46 Term coined by Servigne and Stevens, *How everything can collapse.*

47 Servigne and Stevens, *How everything can collapse.*

48 This view is widespread among many earth scientists, politicians and the general public: See for example: Evan Morison *et al.*, *Climate of the nation 2024: tracking Australia's attitudes towards climate change and energy* (Canberra: The Australia Institute, 2024).

49 Such is the seriousness of our predicament that former NASA scientists and colleagues now argue that the pace and scale of climate breakdown is such that emissions reductions are not enough, and that geoengineering measures are necessary: James E. Hansen, Makiko Sato, Leon Simons, Larissa S. Nazarenko, Inderpreet Sangha, Pushker Kharecha, James C. Zachos, Karina von Schuckmann, Norman G. Loeb, Marlene B. Osman, Qinjian Jin,

When collapse is acknowledged, the response is invariably framed in binary terms: hope versus despair, pessimism versus optimism, order versus chaos. Some commentators continue to insist on geoengineered solutions which are largely unproven, hypothetical and sometimes entirely fantastical – while others dismiss the idea of collapse altogether, accusing those who raise the alarm of being 'doomsters', 'alarmists', 'naysayers' and 'catastrophizers'.[50]

Such derogatory labels however, as we shall see in the next chapter, have not deterred growing numbers of scientists from speaking out, or various activist organisations from highlighting the dangers ahead. Activist groups like Extinction Rebellion, Fridays for Future, Blockade Australia, Climate Action Network Australia, Australian Youth Climate Coalition, Bushfire Survivors for Climate Action, among many others, understand the magnitude of the dangers we face.[51] Thousands of court cases instigated by climate activists have pushed the cause of climate justice,[52] while direct actions have disrupted industries and forced political conversations.[53] Shareholders have pressured corporations to pull away from investments in fossil fuels.[54] But these efforts – frequently spearheaded by a passionate minority – are often met with derision, serious legal consequences, and public hostility. Moreover, in recent years, governments have imposed heavier

Giorgi Tselioudis, Eunjee Jeong, Andrew Lacis, Reto Ruedy, Gary Russell, Jingyi Cao and Jiang Li, 'Global warming in the pipeline', *Oxford Open Climate Change* 3/1 (2023).

50 See for example: Ketan Joshi, 'Climate fatalism is just as wrong as climate denial', *Crikey* (14 July 2025), <https://www.crikey.com.au/2025/07/14/climate-fatalism-as-bad-as-climate-denial>, accessed 12 October 2025.

51 As the direct-action group Blockade Australia states on its website: 'Australia's political and economic systems, and others like it globally, are facilitating the collapse of the ecosystems that support all human and non-human life'. See: Blockade Australia, Blockade Australia website (no date), <https://www.blockadeaustralia.com/>, accessed 12 March 2025.

52 Hannah Otte, 'The lawsuits that could save the world: inside climate courtroom battles', *Climate Cosmos* (13 June 2025), <https://climatecosmos.com/climate-news/the-lawsuits-that-could-save-the-world-inside-climate-courtroom-battles/>, accessed 12 March 2025.

53 One of the most effective groups in this regard has been School Strike 4 Climate, which since 2018 has organised many large-scale national rallies attracting considerable public attention. See the organisation's website: School Strike 4 Climate, *School Strike 4 Climate website* (no date), <https://www.schoolstrike4climate.com/>, accessed 12 March 2025.

54 Mikayla Malinas, Hannah Biggins, Tessa Hall, Daisy Banyasz and Varun Paras, 'Activist shareholders are becoming more efficient, more sophisticated and better resourced', *Allens* (23 September 2024), <https://www.allens.com.au/insights-news/insights/2024/09/activist-shareholders-are-becoming-more-efficient-more-sophisticated-and-better-resourced/>, accessed 12 March 2025.

penalties on those who take the struggle for climate justice to the streets or the headquarters of fossil fuel companies.[55]

Such actions reveal the fault lines emerging between governments, corporations and civil society – tensions that are surely the harbingers of conflicts to come. The 2020 COVID-19 pandemic illustrated just how quickly dominant regimes can be undermined. Borders were shut down. Supply chains faltered. Governments overreached. Entire industries collapsed. There were fistfights in supermarkets, mass protests, civil disobedience, and violent encounters between police and protesters.[56] Some citizens were denied re-entry to their own country. For Australians, this crisis gave rise to new, unexpected anxieties.

If state power and civil society reacted this way to a pandemic, what then might eventuate when environmental disasters escalate and communities further splinter? We are perhaps seeing the answer to this in many of the draconian measures now being adopted by liberal democracies like the United States, United Kingdom and Australia, as well as in more authoritarian regimes where human rights have been 'suspended' or done away with altogether.[57]

The world of course did not end in the wake of the pandemic – but it shifted, exposing deep social and political divisions that turned many lives upside down. Similar shifts and changes will inevitably ensue in the wake of deteriorating biospheric conditions. Consider for a moment what societies might look like in the midst of what Elizabeth Kolbert describes as the 'sixth extinction',[58] or what American journalist, David Wallace-Wells's has referred to as an 'uninhabitable Earth'.[59] If, as Australian essayist James Bradley argues, we are being slowly but surely 'unearthed', then we should not expect powerful elites to come to our collective rescue.[60] Put bluntly: they do not want the climate catastrophe to distract us from the usual business of consumption and profit-making.

---

55 See: Aisha Sharma, 'Australia's criminalisation of peaceful protest is a stain on our democracy', *Crikey* (24 July 2024), <https://www.crikey.com.au/2024/07/24/climate-protest-criminalised-australia/>, accessed 12 March 2025

56 For example, see: Dean Keane, 'Protest penalties, including jail time and $50k fines, prompt backlash from human rights groups', *ABC News* (19 May 2023), <https://www.abc.net.au/news/2023-05-19/unions-and-human-rights-groups-slam-sa-protest-law-change/102367076>, accessed 12 March 2025.

57 Amnesty International, *The state of the world's human rights: Amnesty International's annual report 2024/25* (29 April 2025), <https://www.amnesty.org/en/latest/news/2025/04/global-human-rights-crisis-trump-effect-accelerates-destructive-trends/>, accessed 12 March 2025.

58 Elizabeth Kolbert, *The sixth extinction: an unnatural history* (New York: Henry Holt and Company, 2014).

59 David Wallace-Wells, *The uninhabitable earth: life after warming* (London: Penguin Books, 2019).

60 James Bradley, 'Unearthed', *Meanjin* 78/3 (Spring 2019).

But this form of radical nihilism is much worse than we might think. In a May 2025 article titled 'End Times Fascism', Naomi Klein and Astra Taylor draw attention to the nihilistic tendencies of corporate elites who, they claim, are knowingly taking the planet to brink of destruction while planning their own means of escape. The rest of us, it seems, will have to fend for ourselves.[61] The backstory to this horrifying scenario, say Klein and Taylor, is existential threat, '... from climate breakdown to nuclear war to skyrocketing inequality and unregulated AI – but contemporary far-right – movements financially and ideologically committed to deepening those threats, lack any credible vision for a hopeful future. The average voter is offered only remixes of a bygone past, alongside the sadistic pleasures of dominance over an ever-expanding assemblage of dehumanized others'. These stories are being peddled despite the billionaire class knowing, 'full well that we are entering an age of emergency, but have responded by embracing lethal yet self-serving delusions. Having bought into various apartheid fantasies of bunkered safety, they are choosing to let the Earth burn'.[62]

Much of the opposition to the climate science over recent times has come from this quarter, ably assisted by errant politicians and the corporate media.[63] The fossil fuel industry and its enablers, particularly 'the merchants of doubt', have made life difficult for scientists, activists and concerned citizens alike,[64] making open, transparent discussion difficult.

The alarm bells, of course, were rung years ago, as early as the 1950s, with warnings of the consequences of inaction on greenhouse gases.[65] Global warming resulting from fossil fuel emissions was first identified in the late nineteenth century by Swedish scientist Svante Arrhenius (1859–1927).[66] Since then, numerous

---

61 Naomi Klein and Astra Taylor, 'The rise of end times fascism', *The Guardian* (13 April 2025), <https://www.theguardian.com/us-news/ng-interactive/2025/apr/13/end-times-fascism-far-right-trump-musk>, accessed 12 October 2055.

62 Klein and Taylor, 'The rise of end times fascism'.

63 In addition to a number of corporate media outlets, climate disinformation has been peddled by numerous influential podcasters who have enormous reach. For analysis, see: Oliver Milman, 'Joe Rogan claims study shows Earth cooling – but report's authors say he's wrong', *The Guardian* (4 September 2025), <https://www.theguardian.com/environment/2025/sep/04/joe-rogan-climate-cooling-misinformation>, accessed 05 September 2025.

64 Naomi Oreskes and Erik M. Conway, Merchants of doubt: how a handful of scientists obscured the truth on issues from tobacco smoke to climate change (New York: Bloomsbury Press, 2010).

65 Patrick Pester, 'When did scientists first warn humanity about climate change?', *Live Science* (12 December 2021), <https://www.livescience.com/humans-first-warned-about-climate-change>, accessed 12 March 2025.

66 Stephen Graham, 'Svante Arrhenius', *NASA Earth Observatory* (18 January 2000), <https://earthobservatory.nasa.gov/features/Arrhenius>, accessed 12 March 2025.

scientific studies have confirmed the gravity of our planetary situation – made abundantly clear at many Conference of Parties (COP) and other gatherings.

Although not fully aware of the full scale of the unfolding catastrophe, more and more people in Australia and elsewhere now accept there is a major environmental problem, made worse by rising emissions, feedback loops, multiplier effects, and exponential trends.[67] The question now is how can civil society build the collective capacity required to respond to what is unfolding.

Rethinking everyday civic life

The fracturing of old certainties – stories of growth and progress – will increasingly shape how we think and act. It is in the liminal space straddling the rupture of the old order and the unknown that pressing ontological questions are beginning to arise – about life, death, meaning and purpose. Such questions may help guide us through difficult times. Concerns over justice, equity, and human rights will sharpen as communities confront the breakdown of systems they once relied upon. If Australians are to build the resilience needed to navigate a more hazardous world, they will have to reconnect and participate with others in mutual respect and civic purpose.

Accepting that some sort of liveable future will require the creation of more civic-minded cultures is key to how we respond. Intuitively, we know this already. After all, when disasters strike, it is local people that take the lead. As Indian environmentalist scholar and activist Vandana Shiva points out: 'Civil society is the connective tissue that holds the social body together – especially when institutions falter. In the face of climate collapse, it is often communities, not governments, that first respond, resist, and reimagine'.[68]

New conversations must therefore evolve across social divides, fostering solidarity and cooperation. This is a call for collective responsibility: without a just and coordinated response, we risk sliding into a violent and exclusionary future. The best chance we might have is to invest in localised actions that promote social

---

67 See: Michael Franco, 'Feedback loops are overlooked engines of climate change, says study', *New Atlas* (17 February 2023), <https://newatlas.com/environment/feedback-loops-climate-change/>, accessed 13 March 2025; United Nations, 'Five ways the climate crisis impacts human security', *United Nations Climate Action* (2025), <https://www.un.org/en/climatechange/science/climate-issues/human-security>, accessed 10 October 2025.

68 Vandana Shiva, *Earth democracy: justice, sustainability, and peace* (Cambridge, MA: South End Press, 2005), 32.

cohesion and, and to reconstitute the idea of the commons – a sense of solidarity based on trust, reciprocity and shared norms.[69]

For many climate activists such conversations are deeply troubling. Many are opposed to adaptation responses. Over decades, the activist agenda has been driven by the 'struggle for environmental justice' and the pursuit of 'activist hope'.[70] Protests, marches, petitions, and desperate appeals to the powerful have persisted longer than most of us can remember. These actions have made a difference. Battles have been won. All that is true. And yet, more and more activists have come to believe that the global environmental justice movement has lost the war, in part because of its failure to address the root causes of the climate disaster – extractivist capitalism.[71] The building of community resilience in this context takes on even greater importance. None of this means abandoning activism. Our commitment to changing the world has always been driven by deep moral and ethical convictions. But it is now time to pivot – yes, we should keep up the struggle to reduce emissions, but we also need to recognise that there is no avoiding a hothouse Earth. It is now time to reinvest massively in protective infrastructure and, above all, to build communities capable of withstanding what is to come.

As Jem Bendell' makes clear in *Breaking Together – A freedom loving response to collapse*, embracing our predicament will prompt uncomfortable yet illuminating feelings and conversations.[72] By 'allowing the full pain of our predicament to liberate us' we can, says Bendell, 'live more courageously and creatively'.[73] This is the nature of our predicament: acknowledging life in a time of endings, and seeking rays of light in the darkness. In this sense, collapse is not merely an end, but a passage – an opportunity to rediscover interconnection, creativity, and mutuality as the myths of progress and mastery fade.

Yet any talk of a collective 'we' must acknowledge the deep inequalities that shape peoples' experiences. It's an obvious point, but the onset of climate breakdown and related problems are already having a devastating effect on communities

---

69 Esther Carmen, Ioan Fazey, Helen Ross, Melissa Bedinger, Fiona M. Smith, Katrin Prager, Kerri McClymont and David S. Morrison, 'Building community resilience in a context of climate change: the role of social capital', *Ambio* 51 (2022), 1371–1387.

70 Macy and Johnstone, *Active hope*.

71 See: Klein, *This Changes Everything*.

72 Jem Bendell, *Breaking together: a freedom-loving response to collapse* (Bristol: Good Works, an imprint of the Schumacher Institute, 2023).

73 Bendell, *Breaking together*.

both here and overseas.[74] Brown and Black people, the poor, the homeless, the aged, those with disability or mental health problems will face the most acute challenges as collapse takes its course. Women – still overwhelmingly the primary carers and among the most economically vulnerable – will bear the brunt of these shifts, especially as societal violence intensifies and inequality deepens.[75]

The differentiated nature of environmental harms has long been part of the 'climate change' story. As Naomi Klein argues in her 2007 book *The Shock Doctrine: The Rise of Disaster Capitalism*, those on the economic fringes of society are most at the mercy of rapacious entities, and often than not, ignored by governments.[76] Structural inequalities in western nations run deep. In Australia, for example, young people, aged 18–44 years, make up well over a third of the electorate, yet face some of the worst privations. Locked out of the housing market, paying exorbitant rents and weighed down by educational and other debts, they face the environmental crisis with mounting trepidation and understandable anger at those who let this happen.[77]

Canadian writer and social commentator Brit Wray refers to this age cohort in her 2020 book as *Generation Dread* – a generation terrified of what climate chaos will foist upon them as they are forced to make life-altering decisions, which in Wray's case, was whether or not to have children.[78] Wray did not give up; she decided to have a child. Like other young people, she talks about balancing emotions in the face of the unthinkable, seeking joy and support through 'a strong connection to community that acts together to care for itself'.[79] As collapsologists Pablo Servigne, Raphael Stevens and Gauthier Chapelle argue in *Another End of the World is Possible*, courage, practical skills, and new

---

74 The potential for civilisational collapse as a result of deteriorating climatic conditions and related factors has long been spelt out by climate scientist. See: Moses, 'Collapse of civilisation is the most likely outcome'.

75 Women's Environmental Leadership Australia (WELA), *Gender, climate and environmental justice in Australia: WELA report* (2024), <https://wela.org.au/gender-climate-report/>, accessed 17 March 2025. Among other things, the report shows that women are far more likely to die because of extreme weather events and to be displaced.

76 Naomi Klein, *The shock doctrine: the rise of disaster capitalism* (New York: Metropolitan Books, 2007).

77 Monash Centre for Youth Policy and Education Practice, 'Australian Youth Barometer 2024', *Monash University, Lens* (1 August 2024), <https://lens.monash.edu/@education/2024/08/01/1386898/youth-barometer-paints-a-bleak-picture-of-young-australians-in-2024>, accessed 12 March 2025.

78 Britt Wray, *Generation dread: finding purpose in an age of climate crisis* (Toronto: Knopf Canada, 2022).

79 Wray, *Generation dread*, 232.

frameworks of understanding will be needed to rebuild life within the ruins of industrial modernity. What might have seemed important in the recent past will melt away as assuredly as today's glaciers, or be replaced by different priorities and values. Within the ruins of industrial modernism, we might even talk to each other about the sacred values that make life worthwhile: love, belonging, kinship with all living beings, gratitude, and creative expression. All these and more may rekindle our spirits.

It remains possible, even amidst collapse, to live with meaning and joy. Servigne and colleagues suggest that we may 'live both an apocalypse and a happy collapse'.[80] On a planet stripped of abundance, we must relearn mutual respect, regeneration, and resourcefulness. Practical adaptation, supported by major public investment under new fiscal regimes, must now be the backbone of any societal response. Civil society must push for this, and governments must act decisively – including extracting reparations from fossil fuel corporations. Without such shifts, as independent MP Zali Steggall warns, communities face disaster.

Nothing on Earth is certain – not even human survival. Human life in some form will likely persist, and new life forms may emerge, with or without human presence. Bearing witness to this does not mean we remain passive or disengaged. On the contrary, our predicament calls for renewed solidarity and the creation of life-affirming ways of being.

As this book argues, collapse is an invitation to re-examine how we live with one another and with the more-than-human world. The capacity to act with compassion, humility, and courage becomes central, not as moral ornament, but as modes of survival and renewal. In the face of collapsing systems all of us face a reckoning – mainly with ourselves. As noted in Chapter 4, coming to grips with the new reality will involve considerable inner work as grief and loss take hold. We will need the committed, loving care of others – that is why a genuine sense of community and relational resilience matters. These are the bedrock of life. As we re-engage with each other, new stories will emerge, and we might shed the pernicious idea that humankind is at the centre of life, and that the Earth is here simply for our benefit. As north American Indigenous author, academic and activist Robin Wall Kimmerer remarks in her book, *Braiding Sweetgrass: Indigenous Wisdoms, Scientific Knowledge and the Teachings of Plants*: 'We are bound in a pact of reciprocity, a pact of natural responsibility to sustain those who sustain us …

80 Pablo Servigne, Raphaël Stevens and Gauthier Chapelle, *Another end of the world is possible: living the collapse (and not merely surviving it)* (Cambridge: Polity Press, 2021), 201.

Had we only taken that which is given to us, had we reciprocated the gift, we would not have to fear our own atmosphere today'.[81]

But fear it, we do. The changes needed to live on a hothouse Earth, as we show in the last three chapters, are far-reaching, and may well be beyond the abilities of some. It is difficult to invoke the sort of ontological and systems change we are suggesting if people remain invested in past illusions.

For the majority of us, we believe, there will be little choice other than to change. When the process of societal collapse accelerates there will be confusion and fear as people seek out leaders, fixers, the technocrats, and the doers. Collapse will usher social and political chaos because our lack of general preparedness will leave us exposed. It is unlikely that there will be a neat, ordered transition toward a new order. Many of the musings on climate collapse seem to suggest a sort of prefigurative state that will magically bring us together, thereby avoiding panic and breakdown.[82] It is generally unlikely to work out like that. There is little doubt, for example, that in a situation of confusion, powerful entities will seek to quell signs of dissent and impose their own version of order – a phenomenon referred to by media scholar Hanna Morris as 'apocalyptical authoritarianism'.[83]

Many responders – particularly those who have done some preparatory work – may have a handle on what is happening around them, and what to do. Most, we suspect, will rely on their own resources, and may even shun attempts at building unity with others. It is doubtful too that we will easily bridge rural, regional, metropolitan divides simply by appealing to common interest, although this may work in some quarters. In other words, the response to the process of collapse will be uneven and patchy, and (most worryingly) subject to the vagaries of power and interest. Yet, it doesn't mean we are powerless, in fact, as we explore in the second part of this book, we do have agency.

---

81 Robin Wall Kimmerer, *Braiding sweetgrass: indigenous wisdom, scientific knowledge, and the teachings of plants* (Minneapolis, MN: Milkweed Editions, 2013), 382.

82 While acknowledging the capacity of some communities to respond to disaster cooperatively through mutual aid – see: Rebecca Solnit, *Paradise built in hell: the extraordinary communities that arise in disaster* (New York: Viking Press, 2009) – it is also the case that the opposite can and often does occur. Conflict and discord can indeed be a feature of post-disaster scenarios. See: Australian Institute for Disaster Resilience (AIDR), *Communities in post-disaster recovery: working with conflict in and between* (Melbourne: AIDR, 2020).

83 Hannah Morris, *Apocalyptic authoritarianism: climate crisis, media, and power* (Oxford: Oxford University Press, 2025). Morris shows how climate chaos has been linked to issues of migration, civil unrest and the loss of faith in democracy, thereby ushering increased authoritarian measures to deal with perceived threats to social order – most notably, climate activists. As climate-induced chaos ensures, the more repressive states will likely become.

We can draw inspiration from liberationist writers like Paolo Freire[84] and activist organisers like Matthew Bolton[85] who demonstrate that when people are treated with dignity and respect, when their private stories are connected to a bigger picture, and when they have a sense of shared purpose, then anything is possible.

Conclusion

This chapter has argued that we are living through the terminal phase of a global industrial-capitalist system – a system that has defined modernity's logic of progress, growth, and domination. Its extractive foundations have yielded extraordinary material gains for humanity, yet at the cost of planetary destabilisation, mass extinction, and social fragmentation. What was once framed as the triumph of human ingenuity now reveals itself as self-destructive system that has exhausted both its ecological wonders and moral legitimacy.

To acknowledge this is not to surrender to despair, but to name reality with honesty and clarity. The collapse of industrial modernity is not a single apocalyptic event, but a cumulative process already underway, marked by intensifying climatic disruptions, eroding institutions, and fraying social life. As this chapter has shown, making sense of this predicament is difficult not only because of its scale and complexity, but also because it unsettles the very epistemic, cultural, and affective foundations of modern life. The climate catastrophe is therefore not just a scientific or political predicament; it is a crisis of meaning, testing the adequacy of our languages, institutions, and imaginations.

Breaking the silence that surrounds this predicament is both an ethical and civic imperative. It demands that we move beyond denial and naïve technocratic optimism, and begin to cultivate more honest, inclusive and grounded conversations about what it means to live amid collapse. This involves recognising that the work ahead is not solely about mitigation or adaptation, but reconstituting our social and moral worlds. Building new forms of solidarity – rooted in reciprocity, mutual aid, and ecological interdependence – becomes essential to any viable response.

In the next chapter, we turn to the scientific evidence on the climate disaster to examine how anthropocentric meddling has upended Earth's fragile ecological systems. Understanding the physical processes of the climate predicament – its causes, feedbacks, thresholds, and trajectories – is essential to reflect on how we collectively respond, rather than react, to what is unfolding.

---

84 Paulo Freire, *Pedagogy of the oppressed* (New York: Herder and Herder, 1970).

85 Matthew Bolton, *How to resist: turn protest to power* (London: Bloomsbury Publishing, 2017).

The trauma, the science, the unravelling

> While those of us working in the climate science field know the true picture, and understand the implications for our world, most others do not. And this is a problem – a big one. After all, we can't act effectively to tackle a crisis if we don't know its full depth and extent.
>
> —BILL MAGUIRE

> As a society we are caught between a sense of impending apocalypse and the fear of acknowledging it.
>
> —JOANNA MACY

Introduction

To speak honestly about climate breakdown today is to confront a reality that is deeply troubling. The data are clear, yet their implications remain difficult to absorb. Behind the graphs, projections, and acronyms lies a deeper human story – one of grief, disbelief, and struggle to make sense of an unfolding catastrophe. For those who work closest to the evidence – scientists, ecologists, and allied researchers – their task has evolved beyond simply accumulating knowledge, to become an act of personal sensemaking. They witness not only the degradation of fragile ecosystems but also the dissolving of old certainties about progress, stability, and the future itself.

This chapter examines what it means to live and work within that knowledge framework. It explores how the scientific community, having warned for decades of the consequences of continued inaction on greenhouse gases, now finds itself both messenger and mourner in an era of planetary breakdown. The emotional and existential burden carried by them reflects a collective psychological condition – a form of trauma that arises from prolonged exposure to deteriorating environment conditions and political inertia. Their trauma – a profound sense of woundedness – is shared by countless others who witness the horrors of ecological loss and the continuation of a system seemingly hell-bent on self-destruction.

Here, we chart the contours of ecological collapse through scientific evidence and the lived accounts of those closest to the data – individuals who may be most attuned to the escalating threats faced by Australian communities.

We should listen to them.

'Bumbling along against the current' – Jane's journey

Neither of us – the authors of this book – have met directly with Jane.[1] A close friend with a strong interest in the climate science suggested that, as a coral reef ecologist, Jane might be a good person to talk to about the impacts of climate disruption. So, after obtaining her mobile number, one of us called Jane. It was clear from the outset that she is an extremely intelligent, diligent, warm and generous person – someone who cares deeply and passionately about what's happening to our planet. Thoughtfulness and kindness accompany every word, as does more than a wisp of sadness and frustration about the lack of political action on environmental issues.

We told Jane about one of the main arguments in this book: namely, that based on the science and current trends, we held a profoundly pessimistic outlook (which included our acceptance that we are witnessing a great unravelling), but that we are also very interested in how – as citizens and community members – we can set about building resilient civic cultures capable of responding, rather than simply reacting, to the many existential challenges ahead. None of this seemed to phase Jane – and she agreed that communities need to prepare for what is coming.

For decades, Jane has focused her research on the conservation of ocean reefs, reef recovery, acidification, deteriorating water quality due to the run-off from agriculture, and the limitations to reef recovery posed by climate breakdown. She says that although the loss of biodiversity is deeply concerning, 'climate change' is having the greatest impacts on our ecologies. Among other things, '*it is producing fiercer marine heatwaves, leading to coral bleaching and coral death. But also intensifying storms, flooding, seawater acidification, and so forth*'. Daily scrutiny of the datasets and the frequency of disturbances and degradation to some of the world's most spectacularly beautiful marine ecologies has been enormously difficult for Jane. '*Witnessing mass coral bleaching*', she says,

---

1 Not her real name.

> *has had a significant impact on my mental health and that of many of my colleagues. During bleaching events, we can almost smell the death of the corals. It's so hard to witness this. There are times when I've asked: What's the point? We know how bad things can get; we know how to fix this problem. It's a big challenge. We feel so angry and frustrated at the inaction and the greed causing all this destruction. We've done our best to report on this, to bring the data to the attention of decision makers. But still, most ignore it, or turn a blind eye.*

For Jane, witnessing all this has been traumatic. '*These experiences*', she says,

> *have left me feeling burnt out, exhausted. I was suffering climate grief and anxiety – like a lot of other people. I took nine months off. I just needed to stop doing what I was doing. As time has gone on, I've started feeling a bit more grounded. We are still observing remaining resilience in the reef systems, as most reefs start to recover a few years after having been hit.*

But future-gazing is something Jane finds difficult. '*You can't look too far into the future*' she remarks, '*honestly, I find it difficult to visualize the longer-term future for our ecosystems and even for humanity, the changes are so rapid and there are tipping points*'.

We came back to the vexed matter of how this sense of uncertainty made her feel.

> *What do I feel? Well, immense frustration because all this is preventable. I'm angry at all the greed and inaction. I've done a lot of work to regain a sense of some stability. We're now turning our attention to doing research on solutions. We can't afford not to try to help reefs, to be ready to do ecosystem restoration – but only if we will come to our senses and start serious climate action.*

Jane seeks solace in community activities, immersing herself in nature, and working with students, although all this has a downside too:

> *I'm involved in community; I connect with nature whenever I can, even though it's a constant reminder of what is occurring. I also work closely with my students. It's really important to me. We're seeing loads of students with severe mental health problems. So many are really struggling, lack resilience, feel lonely, isolated. There's no real sense of community for many of them. The destruction of our environment, is everywhere, it's in your face. As scientists we see the evidence all round. I see this all the time.*

For Jane, life has become more uncertain and deeply troubling:

> *It feels like I'm bumbling along against the current of rapid change. It's like we're being sucked down the gurgler. But we can put the plug back in – we could have done it a long time ago. Sure, in the longer-term things will bounce back, some things, but we should care about what we do today and for the next several generations. We've lost our sense of spiritual connection with the world around us. We need more compassion for the Earth.*

Jane's experience is not isolated. Across the world, scientists – and people at large – who have dedicated their lives to understanding and protecting the planet are confronting similar emotional and existential challenges. Grief, anger, and fatigue have become ubiquitous responses. Some have lost hope. To grasp the scale of this suffering is to understand that climate disruption is not only an ecological predicament, but also a profoundly human experience that has, in many cases, upended assumptions of certainty and security.

Climate scientists like Jane – those at the forefront of the knowledge about ecological collapse – have faced enormous challenges, including manufactured doubt, and personal attacks on their professionalism and integrity.[2] Despite their best efforts, these scientists have long struggled to get the seriousness of their findings across to an often sceptical and distracted public. They have also experienced pushback from former 'believers' who now see the 'climate agenda' as an orchestrated attempt to increase governmental control over populations, or to push a corporate 'green agenda'.[3]

For scientists trying to navigate their way through this thicket, the full grim reality of climate breakdown can quickly lose its public resonance. Often they are compelled to address the tension of having to present highly qualified data – it is the accepted practice of scientists to be cautious – and encouraging people to accept what is being presented to them.[4] Until fairly recently, the language deployed by climate researchers – at least in public – has been tempered by fears that they might be accused of deliberately exaggerating or distorting their findings for public effect; that they are beholden to certain vested interests, or that they have abandoned objectivity in pursuit of individual gain.[5]

---

2 See: Michael Mann and Peter Hotez, 'Science is under siege from weaponised disinformation – posing a threat to human civilisation' (9 September 2025), *Australian Online News*, <https://australianonlinenews.com.au/2025/09/09/science-is-under-siege-from-weaponised-disinformation-posing-a-threat-to-human-civilisation-michael-mann-and-peter-hotez/>, accessed 4 October 2025.

3 See for example: Matt Binder, 'Who really profits from climate fear – the billionaires behind the green agenda' (28 January 2025), *Climate Cosmos*, <https://climatecosmos.com/climate-society/who-really-profits-from-climate-fear-the-billionaires-behind-the-green-agenda/>, accessed 17 March 2025.

4 See: Naomi Oreskes, *Why trust science?* (Princeton: Princeton University Press, 2018).

5 Christel W. van Eck and Toni G. L. A. van der Meer, 'Narratives of hope and concern? examining the impact of climate scientists' communication on credibility and engagement', *Public understanding of science*, 34/6 (2025), 734–51, <https://doi.org/10.1177/09636625251314159>, accessed 17 March 2025.

These are deeply wounding accusations, which impact what is said, and how it is said.[6] Despite such obstacles, growing numbers of scientists have ventured forth into the public arena to air their feelings about what they are witnessing.

The trauma shadow

How do those closest to the climate data think and feel about what they are witnessing? Before answering this, it is important to recognise that emotional distress in response to ecological disruption is both understandable and widespread – including for the authors and readers of this book. Earth system scientists are routinely exposed to the realities of a transforming climate. Their work is conducted within the parameters of clearly defined professional ethics and peer-reviewed scholarly inquiry. Despite frequent attempts to pull the science apart, to undermine its legitimacy and impugn the reparations of scientists,[7] the information gathered through decades of carefully conducted research has elicited a broad consensus on the environmental challenges we now face.

While there are disputes over the pace and scale of what is occurring, and a few scientists deny that there is a problem at all,[8] the vast majority now accept that the rapid deterioration in the Earth's ecological systems can be linked directly to human activity.[9] The accumulated evidence is both dauntingly persuasive and deeply troubling. How could it be otherwise, when it points clearly to the rapid acceleration of changes in the Earth's fragile planetary ecosystems? It is no exaggeration to describe the emotional response to such knowledge as traumatic

---

6 Nancy Shute, 'When attacks on science threaten our survival', *Science News*, 2 May 2021, <https://www.sciencenews.org/article/when-attacks-on-science-threaten-our-survival>.

7 Doug Hilton, 'We need science, and we need to defend it from attack' (18 April 2024), *CSIRO News*, <https://www.csiro.au/en/news/all/articles/2024/april/defending-science>, accessed 17 March 2025; Damian Carrington, 'Climate misinformation turning crisis into catastrophe, report says' (19 June 2025), *The Guardian*, <https://www.theguardian.com/environment/2025/jun/19/climate-misinformation-turning-crisis-into-catastrophe-ipie-report>, accessed 17 March 2025.

8 John Cook, Dana Nuccitelli, Sarah A. Green, Mark Richardson, Bärbel Winkler, Rob Painting, Robert Way, Peter Jacobs and Andrew Skuce, 'Quantifying the consensus on anthropogenic global warming in the scientific literature', *Environmental research letters* 8/2 (2013), 024024, available at <https://iopscience.iop.org/article/10.1088/1748-9326/8/2/024024> (accessed 18 March 2025).

9 Mark Lynas, Benjamin Z. Houlton and Simon Perry, 'Greater than 99% consensus on human caused climate change in the peer-reviewed scientific literature', *Environmental research letters* 16/11 (2021), 114005, available at <https://iopscience.iop.org/article/10.1088/1748-9326/ac2966> (accessed 18 March 2025).

– an understandable reaction to what is unfolding. Living with the knowledge of what a hothouse Earth means for present and future generations is, to say the least, profoundly unsettling. It is part of a collective trauma – one that unsettles identity, belonging, and purpose on a civilisational scale.

Trauma theorists such as Peter Levine, Sandra Bloom, Gabor Maté, Judith Cohen, Bessel van der Kolk, and others view trauma as a fundamental shift in how we experience the world following a deeply troubling event or series of events.[10] In *The Myth of Normal: Trauma, Illness and Healing in a Toxic Culture*, Maté argues that in the wake of such personal upheaval, a sense of woundedness, 'dictates much of our behaviour, shapes our social habits, and informs our ways of thinking about the world'.[11]

According to Maté, the effects of trauma are exhibited in our emotional responses and routine interactions with others. At worst, it can give rise to deep and enduring unhappiness, disassociation and disengagement with the social world. Its effects on body, mind and spirit – as Maté knows from his own childhood experience – are profound, often unaddressed, and long-lasting.[12] While many trauma reactions originate in childhood, they can arise at any point in the lifecycle, often through the repetition of triggering experiences and the accumulation of distressing knowledge. Vicarious trauma – the unwanted by-product of bearing witness to another's anguish – is as real in its emotional, physical, and behavioural intensity as if experienced directly.[13] To be repeatedly exposed to harrowing climate data and to bear witness to more and more extreme weather events is to trigger the darkest and most troubling of thoughts and feelings.

---

10 There is an abundance of trauma theory – a field of inquiry that has mushroomed over recent years. See: Nicole A. Sütterlin, 'History of trauma theory', in *The Routledge Companion to Literature and Trauma*, ed. by Colin Davis and Hanna Meretoja (Abingdon: Routledge, 2020), pp. 11–22**, available at <https://www.taylorfrancis.com/chapters/edit/10.4324/9781351025225-1/history-trauma-theory-nicole-s%C3%BCtterlin> (accessed 18 March 2025).

11 Gabor Maté and Daniel Maté, *The myth of normal: trauma, illness, and healing in a toxic culture* (New York: Avery, 2022).

12 Kirsty Young, 'Gabor Maté: Seven things we learned when he spoke to Kirsty Young', *BBC Young Again* [no date], available at <https://www.bbc.co.uk/programmes/articles/3TdSk6dlD-mFwVYFzB34vXZZ/gabor-mate-seven-things-we-learned-when-he-spoke-to-kirsty-young> (accessed 18 March 2025).

13 Lisa A. McCann and Laurie Anne Pearlman, 'Vicarious traumatisation: a framework for understanding the psychological effects of working with victims', *Journal of traumatic stress* 3/1 (1990), 131–49, available at <https://link.springer.com/article/10.1007/BF00975140> (accessed 18 March 2025).

Such responses – which can long nestle in the subconscious – are not easily shaken off or forgotten. They sit in the accumulated store of memory and are expressed through a range of somatic effects, often compounding previous traumatic experiences. Convenient taxonomic labels like 'eco-anxiety', 'eco-grief' and 'solastalgia', while signalling the depths of psychological responses to climate chaos, somehow fail to capture the lived experience of scientists – and anyone in fact who engages closely with the realities of a rapidly changing biosphere. Medicalising these reactions can also ignore the collective and political dimensions of climate distress.[14]

The enormity of what faces us – a truly existential predicament – and the sense of overwhelm that comes with it, almost inevitably triggers a trauma response. After all, accepting the destruction of Earth's ecologies – indeed, an entire way of life – or entertaining the possibility of societal collapse – is challenging in the extreme. So too is reckoning with the uncertainties and unknowns that such upheavals present.

This burden of knowledge is not confined to climate scientists alone. Recent survey data suggest that this sense of looming disruption is not confined to scientists or activists but is increasingly embedded within the broader Australian public. A nationally representative study of nearly 2,000 Australians found that around 60% are at least moderately concerned about the prospect of social and economic collapse within the next 25 years, with nearly one-third expressing high levels of concern. These anxieties extend beyond abstract environmental change to fears about food shortages, supply chain disruptions, climate refugees, and regional conflict.[15] What is striking is not only the scale of concern, but its normalisation: the possibility of societal breakdown, once considered fringe, is now part of mainstream public consciousness. Yet, as the study notes, this growing awareness remains largely unspoken in public discourse, reinforcing the sense that we are collectively carrying knowledge that is both widely shared and difficult to articulate.

Increasingly, ordinary people – citizens, students, parents, and communities – are carrying a form of what might be called *collective dark knowledge*: the awareness of an impending and pervasive doom for which no adequate social or political response exists. To know, and to feel, that the systems underpinning our

---

[14] This agreement has been persuasively put by: Panu Pihkala, 'Anxiety and the ecological crisis: an analysis of eco-anxiety and climate anxiety', *Sustainability* 12/19 (2020), 7836, available at <https://www.mdpi.com/2071-1050/12/19/7836> (accessed 18 March 2025).

[15] Clive Hamilton, Climate change and social collapse, Research paper 7, Climate Adaptation Project, Charles Sturt University, December 2025, p. 1.

lives are unravelling produces a quiet but profound dissonance between what we understand intellectually and what we can bear emotionally. Many retreat into silence, humour, or distraction, not because they are indifferent, but because speaking openly about ecological collapse threatens to unmoor the fragile sense of normalcy that daily life requires. In this way, trauma becomes both personal and cultural – a condition that shapes what can be said, imagined, or even felt about the future. The question that emerges is not only *what do we know*, but *what do we do with what we know*? How can societies metabolise truths that are almost too large to face, yet too urgent to ignore?

Scientists, like the rest of us who accept empirical evidence, increasingly inhabit a world detached from past certainties. The data flashing across their computer screens is both confirmation of an evolving catastrophe and a portal into an unknowable future. This is the unsettling space occupied by those tasked with tracking ecological collapse. For scientists already apprehensive about the strong public reactions their warnings may provoke – especially given their frequent dismissal as doom-mongers and catastrophisers – speaking openly about the threats we face is profoundly troubling. But over recent years, this has begun to change.

Speaking out

Lead IPCC scientist and author of books such as *Sunburnt Country: The History and Future of Climate Change in Australia* and *Humanity's Moment: A Climate Scientist Case for Hope*, Joëlle Gergis reflects on her years immersed in climate science, detailing not only the empirical realities of ecological collapse but also the profound emotional and spiritual toll of bearing witness to it – compounded by the persistent absence of decisive political action.[16] In a 2022 article in the online journal, *The Conversation*, Gergis reflects on how she has, 'spent hundreds of hours trawling through countless UN reports and scientific papers until my eyes sting and I can no longer absorb any more information. I feel overwhelmed and saturated with sorrow'.[17]

'There are days', she says, 'when it's hard to take in all of the senseless destruction and somehow try and accept so much avoidable suffering'. It is this wanton

[16] Joëlle Gergis, *Humanity's moment: a climate scientist's case for hope* (Melbourne: Black Inc., 2022).

[17] Joëlle Gergis, 'Friday essay: 'I feel my heart breaking today' – a climate scientist's path through grief towards hope', *The Conversation*, 26 August 2022, available at <https://theconversation.com/friday-essay-i-feel-my-heart-breaking-today-a-climate-scientists-path-through-grief-towards-hope-188704> (accessed 18 March 2025).

destruction and political inaction that really distresses Gergis. 'I know that time is running out. I've come to realize that the only way forward is not a detour through denial, but straight through the heartland of grief'. Confronted by the full horror of what she is witnessing, the emotional dam threatens to burst open: 'When you realize that all that sustains us is at stake, it's almost too much to process'. For Gergis, 'life hangs in the balance', and it's why, 'many scientists also feel angry, scared and sad. We don't often talk about it openly, but that doesn't mean we don't feel really upset about what's going on'.[18]

In a May 2024 article, environmental journalist Damien Carrington reported on interviews with 380 scientists from around the world.[19] He asked about their feelings when confronted with the scientific evidence on climate breakdown. Most talked of their struggles with mental health – depression, anxiety, sleeplessness – as well as their anger and frustration over the lack of global action on heat-trapping gases. Over three-quarters of them said they expected global temperatures to soon exceed the 1.5-degree threshold set under the 2015 United Nations Framework Convention on Climate Change Paris Agreement. (This threshold was indeed crossed in 2024).[20] Worryingly, over half of respondents expected global temperatures to reach 3 degrees or more by the end of the century – a truly terrifying prospect. Words and phrases like 'infuriating', 'distressing', 'overwhelming', 'despair', 'guilt', 'scared mightily', 'fear', 'resigned to disaster' reflected the troubling emotions experienced by these scientists. Anger too lurked near the surface, particularly with regard to the complicity and duplicity of the fossil fuel corporations and governments who, according to these scientists, have refused to act with the required moral urgency, even in the face of the most alarming evidence.

The majority of those interviewed by Carrington have virtually given up on saving the planet, believing that their work has been ignored. As one respondent put it: 'Scientists are human: we are also people living on this Earth, who are also experiencing the impacts of climate change, who also have children, and who also have worries about the future… We did our science, we put this really good report together and – wow – it really didn't make a difference on the policy. It's

---

18 Gergis, 'Friday essay: I feel my heart breaking today'.

19 Damian Carrington, 'World's top climate scientists expect global heating to blast past 1.5c target', *The Guardian*, 8 May 2024, available at <https://www.theguardian.com/environment/2024/may/08/worlds-top-climate-scientists-expect-global-heating-to-blast-past-1-5c-target> (accessed 18 March 2025).

20 Tyne Logan, 'In 2024 global warming topped 1.5c: have we failed the Paris Agreement?', *ABC News*, 13 January 2024, available at <https://www.abc.net.au/news/2025-01-13/a-year-above-1-5c-what-does-it-really-mean/104774364> (accessed 18 March 2025).

very difficult to see that, every time'. A few still cling to optimism, desperately hoping against hope that governments might finally come to the rescue, even though the prospects of doing so seems remote. Generally, the mood among this group was bleak and despondent – their take on the future full of spine-chilling concern. As one person put it: 'Humanity is heading towards destruction'.[21]

The sense of barely concealed despair is borne out in many other studies. For example, in 2021, one of us – Jean Renouf, then university lecturer and researcher – reported on qualitative interviews conducted with sixteen Earth scientists from twelve countries.[22] Thirteen of them had at various times contributed to IPCC reports. Renouf observed how awareness of the planet's predicament, and the failure of governments to take urgent action, had percolated into every aspect of the scientists' personal and professional lives. Despite some differences in their emotional responses – much depended on individual temperaments and support systems – Renouf found that, 'climate experts almost always viewed climate change through the prism of its impacts. They see it as a disruption of the normal way of life and all concur about the gravity of the situation'.[23] Many however, clung to the wavering hope that political action might ultimately prevent the worst – that is, runaway climate disaster. Others were on the edge of total despair. (It's worth noting that the interviews for this study were conducted in 2019. Since then, extreme weather events have accelerated and intensified).[24]

Many climate experts, it seems, deal with the threat of catastrophe by building an emotional shield around their professional work. Generally, their emotional responses depended on factors such as family environment, the nature of work undertaken, and the effects of witnessing extreme weather events. Most of Renouf's participants focused on food and water shortages as the most pressing danger facing humankind. They predicted widespread chaos and outbreaks of violent disorder. One respondent remarked: 'Within the lifetime of most people around us, we will see huge disruptions. We will see issues with food security, but more importantly we will just see a lot of violence. That's what people tend to do when they get upset and hungry, and it's not going to be pretty'. 'I worry about

---

21 Carrington, 'World's top climate scientists', n.p.

22 Jean S. Renouf, 'Making sense of climate change: the lived experience of experts', *Climatic change* 164 (2021), available at <https://doi.org/10.1007/s10584-021-02986-5> (accessed 18 March 2025).

23 Renouf, 'Making sense of climate change', 14.

24 James E. Hansen, Pushker Kharecha, Makiko Sato, George Tselioudis, Jon Kelly, Susanne E. Bauer… Auli Pokela, 'Global warming has accelerated: are the United Nations and the public well-informed?', *Environment: science and policy for sustainable development* 67/1 (2025), 6–44, available at <https://doi.org/10.1080/00139157.2025.2434494> (accessed 18 March 2025).

the world I will be leaving to my grandchildren' said another, 'because I don't think we've done nearly enough to secure their future… well, it's something I regret. It's certainly something I feel bad about'.[25] Are these scientists experiencing some form of trauma? And if they are, what does this mean for them, for their work, their personal relationships? Clearly, they get on with their lives, they go to work, they enjoy many of the things life has to offer. But they do so in the shadow of trauma, the wounding knowledge that the world as they know it is changing beyond recognition and that there is no going back. The destruction of the planet's beauty, wonder and life-sustaining systems, weighs heavily on them.

Renouf's interviewees were perhaps most fearful of the gathering pace of climate breakdown, and the societal collapse it threatens. But they were also experiencing grief and loss over the loss of the sense of stability that once characterised life in the Holocene. If these same scientists were interviewed today, six years on, they would undoubtedly be significantly more anxious and depressed over the Earth's plight. Hope would be in short supply.

We are reminded here of the subtitle to Joëlle Gergis's book, *A Scientist's Case for Hope* which is yet another plaintive cry for urgent climate action. Interestingly, the tone of Gergis's writings since the publication of her book has changed markedly. There is now a sense of bleakness and barely concealed pessimism in her outlook as the planet lurches from one disaster to another.[26]

Hypocrisy and inaction

Gergis of course, is part of a community of scientists who have repeatedly beseeched governments to take urgent action on the climate. Countless open letters, petitions, public appeals, warnings and expressions of concern, have reverberated through the public arena.[27] But despite all the gatherings – conferences, seminars, COP meetings – there is widespread exasperation at the failure of governments to set meaningful goals and/or to adhere to agreed binding

25 Renouf, 'Making sense of climate change'.

26 This is exhibited in Gergis' s harrowing essay: Joëlle Gergis, 'Highway to hell: climate change and Australia's future', *The Australia Institute*, 28 June 2024, available at <https://australiainstitute.org.au/event/highway_to_hell/> (accessed 18 March 2025).

27 See for example: EcoMind Australia, *Climate science, not climate silence: open letter from scientists and educators* (2025), available at <https://www.ecomind.au/open-letter> (accessed 18 March 2025); The Australia Institute, 'Open letter calls on government to set timeline for end of new fossil fuel projects', 25 September 2024, available at <https://australiainstitute.org.au/post/open-letter-calls-on-government-to-set-timeline-for-end-of-new-fossil-fuel-projects/> (accessed 18 March 2025).

targets.[28] Indeed, many administrations continue to grant licences for new fossil fuel projects – a reality that is truly galling to those at the forefront of the climate science.[29] The contradictions in government policies across the globe – talking up the need for climate action on the one hand, while allowing new greenhouse gas projects to proceed on the other – are there for all to see.

What would Renouf's scientists make, for instance, of a reports by the International Institute for Sustainable Development (IISD) in July 2023 which concluded that, 'Over the past decade, new licences issued by high-capacity, low-dependency countries including the US, the UK, Canada, Australia and Norway are estimated to have contributed five times more greenhouse gas emissions between 2014 and 2023 than all other oil- and gas-producing countries combined'.[30] The US has in effect become the world's leading petrostate, producing around 20 per cent of the world's total output in oil and gas.[31] Growing numbers of licence approvals in the UK, Norway, Canada, and Australia will not only further pollute an already gas-rich atmosphere, leading to even more catastrophic weather events. According to figures provided by the Australian government, as of October 2024, Australia operated 455 major resource and energy projects, an increase from 421 in 2023.[32] Drawing on the IISD's report, journalist Oliver Milman and Nina Lakhani note that: 'The 11.9bn tonnes of greenhouse gas emissions – which is roughly the same as China's annual carbon pollution – resulting over their lifetime from all current and upcoming oil and

---

28 See for example: Daisy Dunne, 'Analysis: 95% of countries miss UN deadline to submit 2035 climate pledges', *Carbon Brief*, 10 February 2025, available at <https://www.carbonbrief.org/analysis-95-of-countries-miss-un-deadline-to-submit-2035-climate-pledges/> (accessed 18 March 2025).

29 See for example the 'carbon bomb' controversy over the Australian Federal Government's approval of an extension to the Woodside North West Shelf gas project: Graham Readfearn, 'Australia's biggest gas project greenlit to 2070 with "partial" protection for Indigenous rock art', *The Guardian*, 12 September 2025, available at <https://www.theguardian.com/environment/2025/sep/12/north-west-shelf-gas-project-extended-to-2070-with-partial-protection-for-indigenous-rock-art> (accessed 18 March 2025).

30 International Institute for Sustainable Development, *Carbon minefields: oil and gas exploration monitor – July edition* (2024), available at <https://www.iisd.org/publications/newsletter/carbon-minefields-monthly-oil-gas-exploration-monitor-july-2024> (accessed 18 March 2025).

31 International Energy Agency, *World energy investment 2025 – United States* (2025), available at <https://www.iea.org/reports/world-energy-investment-2025/united-states> (accessed 18 March 2025).

32 Department of Industry, Science and Resources, *Resources and energy major projects: 2024* (Canberra: Commonwealth of Australia, 2024), available at <https://www.industry.gov.au/publications/resources-and-energy-major-projects-2024> (accessed 18 March 2025).

gas fields forecast to be licensed by the end of 2024 would be greater than the past four years combined'.[33]

While countries like Australia like to portray themselves as clean energy frontrunners, the reality is that they account for a whopping 67 per cent of all new oil and gas licensees issued since 2020. 'The hypocrisy of wealthy nations, historically responsible for the climate crisis, is staggering as they continue to invest heavily in fossil fuels – putting the world on track for unimaginable climate catastrophe while claiming to be climate leaders', says Harjeet Singh, global engagement director for the Fossil Fuel Non-Proliferation Treaty Initiative.[34] This contradiction – between declared commitments to climate action and the continued expansion of fossil fuel extraction – reveals a profound disconnect at the heart of global governance. It exposes how political and economic systems remain structurally tethered to the very forces driving planetary destabilisation.

The record-busting contributions of these nations, understandably pushes many scientists to the brink of despair. As each new dataset reveals the rapidity of ecological breakdown, and as words like 'unprecedented' and 'record-breaking' become commonplace, is it any wonder that some scientists begin to doubt their contributions?

The Great Unravelling

So, what exactly is causing this anguish among climate scientists and other Earth watchers? This might appear a strange question because the answer seems so obvious. A fossil fuel heating planet leads to more extreme weather events, and feedback loops, leading to more destruction of our fragile ecosystems – that, at least is the surface picture. Our intention here is to draw on summaries of some of the most compelling scientific studies that chronicle changing biophysical conditions.

Before doing so, it is worth emphasising another obvious point that the climate disaster is 'only' one of a number of environmental problems impacting the Earth's ecologies. Breaches of what have been termed 'planetary boundaries'

---

33 Oliver Milman and Nina Lakhani, 'Wealthy western countries lead in global oil and gas expansion', *The Guardian*, 25 July 2024, available at <https://www.theguardian.com/environment/article/2024/jul/24/new-oil-gas-emission-data-us-uk> (accessed 18 March 2025).

34 Harjeet Singh, '"They're expanding fossil fuels": Indian activist Harjeet Singh on rich countries', interview for *Democracy Now!*, 21 November 2024, available at <https://www.democracynow.org/2024/11/21/cop_29_finance_agreement_harjeet_singh> (accessed 18 March 2025).

– biosphere changes (biodiversity loss and species extinction), stratospheric ozone depletion, ocean acidification, biogeochemical flows (the transfer of chemicals and simple substances between living systems and the environment), land-system changes (such as deforestation), and freshwater use, are having profound and lasting consequences for life on Earth.[35] Given the enmeshed nature of such systems, a change in one area will inevitably ripple-impact others and result in lasting and even irreversible changes to fragile ecosystems. In 2023, a team of Earth system scientists under the aegis of the Stockholm Resilience Centre at Stockholm University found that apart from ozone depletion and aerosol loading, all boundaries were breached, compromising the health and wellbeing of the planet.[36] Australia has transgressed a number of boundaries, especially in biodiversity, land-system change, and nitrogen and phosphorus flows. It has also exceeded climate boundaries and placed enormous stress on waterways owing to harmful management practices. The key point here, however, is that climate breaches are among the main drivers of environmental changes in other areas, made worse by feedback loops, exponential changes and other compounding factors. Suffice to say, taken together, these breaches pose an existential threat to life on Earth.

Monitoring the climate disaster

One the most authoritative and well-known climate information source is the Intergovernmental Panel on Climate Chage (IPCC). March 2023 saw the publication of the final instalment of an eight-year project focused on the impacts of climate disruption on the Earth's ecological systems.[37] Produced by 782 Earth scientists, the 'AR6' provides a comprehensive peer-reviewed analysis of the nature, scale and impacts of climate breakdown. As with earlier IPCC assessments, the rigour and detail exhibited in its voluminous and data-rich AR6 pages are truly remarkable and indicative of the extreme care and diligence exercised in an often-hostile political atmosphere. That said, IPCC reports have over a number of

---

[35] Katherine Richardson et al., 'Earth beyond six of nine planetary boundaries', *Science advances* 9/37 (2023), eadh2458, available at <https://www.science.org/doi/10.1126/sciadv.adh2458> (accessed 18 March 2025).

[36] Stockholm Resilience Centre, *Planetary boundaries*. 2024. Available at: <https://www.stockholmresilience.org/research/planetary-boundaries.html> (accessed 18 March 2025).

[37] Intergovernmental Panel on Climate Change (IPCC), *Climate change 2023: synthesis report. contribution of Working Groups I, II and III to the Sixth Assessment Report of the Intergovernmental Panel on Climate Change* (2023), available at <https://www.ipcc.ch/report/sixth-assessment-report-cycle/> (accessed 18 March 2025).

years faced considerable criticism, mainly because of occasional errors, conservative bias stemming from over-reliance on consensus (leading to underestimates and dilution of findings), outdated modelling, and the undue influence of member states in the softening of language.[38] If anything, however, these criticisms of the IPCC's approach highlight the fact that the planet is likely in more danger than previously thought.

In *Humanity's Moment*, Joëlle Gergis describes how the IPCC data gathering processes work, highlighting the complex collaborative, multidisciplinary, and international nature of the enterprise, its methodological rigour, the care, and attention to detail, as well as the heavily qualified conclusions.[39]

The AR6 derives from the work of three working groups which recorded the impacts of rising greenhouse gas emissions across the world, describing their destructive effects on ecosystems, agriculture, economies, and entire societies. The report draws attention to the risks we run if, as looks likely, urgent changes are not implemented to reduce heavy concentrations of carbon dioxide, methane and nitrous oxide in the atmosphere. Based on current policies and pledges, if implemented, the report states the world is on track for a temperature increase between 2.0 and 2.7°C by 2100 compared to pre-industrial levels, which is likely to reach tipping points leading to an acceleration of the global heating.[40] As such, the report concludes that:

- *Global temperature rises*, caused in significant part by human activity, have resulted in more frequent and extreme weather events (fires, floods, storms), and rapid melting of sea ice sheets, leading to low-lying coastal and island inundation.

---

38 See for example: Kevin Anderson, 'IPCC's conservative nature masks true scale of action needed to avert catastrophic climate change', *The Conversation*, 24 March 2023, available at <https://theconversation.com/ipccs-conservative-nature-masks-true-scale-of-action-needed-to-avert-catastrophic-climate-change-202287> (accessed 18 March 2025).

39 Gergis, *Humanity's Moment.*

40 On 13 November 2025, *The Guardian* reported on findings by Climate Action Tracker and the Global Carbon Project which concluded that greenhouse gas emissions are on a trajectory that would raise global temperatures by 2.6C degrees above pre-industrial levels by the century's end, and that despite a slowing rate, emersions are still on the rise with a new record high to be hit in 2025. Thus far, national commitments to cut emissions have, say the reports, been well below what is needed globally to reduce catastrophic heating. See: Oliver Milman and Damian Carrington, 'World still on track for catastrophic 2.6c temperature rise, report finds', *The Guardian*, 13 November 2025, available at <https://www.theguardian.com/environment/2025/nov/13/world-still-on-track-catastrophic-2-6c-temperature-rise-report-finds> (accessed 18 March 2025).

- *Every fraction of temperature rise* intensifies already highly destructive climate-related events.
- *Heatwaves* will increase in number, intensity, and duration, leading to more deaths and untold pressure on health and other services, and economies more generally. (In an alarming report published by scientists at the ARC Centre of Excellence for 21st Century Weather and CSIRO, it was concluded that heatwaves will intensify in frequency, length and overall temperature for centuries to come – in some parts of the world for a thousand years – even after the world hits net zero emissions. Compounding these risks is the rise in wet-bulb temperatures – the combined measure of heat and humidity – beyond which the human body can no longer cool itself through sweating. Once this threshold, around 35°C, is crossed even for a few hours, survival becomes impossible without mechanical cooling, rendering large parts of the tropics and subtropics potentially uninhabitable during extreme heat events. The most impacted regions will be around the equator).[41]
- The *bushfires* we saw in Australia in 2019/2020 and across various parts of Europe and North America in 2023–2025 will be more commonplace, leading to even greater concentrations of toxic gases in the atmosphere. Loss of carbon-capturing trees through land-clearing and deforestation have only added to the problem resulting in some forest regions becoming net emitters rather than absorbers of carbon. (James McCarthy and colleagues at the World Resources Institute note that, 'forest fires are becoming more widespread, burning nearly twice as much tree cover today as they did 20 years ago'.[42] In 2025 wildfires ravaged many nations across Europe, Asia and North and South America).
- This *feedback loop effect* also occurs in melting permafrost which, with rising temperatures, will release increasing amounts of methane, a highly potent greenhouse gas. (According to the Climate Council, methane levels are at their highest level in 800,000 years. Over the past 20 years up to 2024, methane has trapped 85 per cent more heat than carbon. Australian agricultural and mining industries emit 4 million tonnes annually).[43]

---

41 University of Melbourne (UoM), 'Scientists warn heatwaves will intensify for 1,000 years even after net zero', *Science Daily*, 20 November 2025, available at <https://www.sciencedaily.com/releases/2025/11/251120002600.htm> (accessed 18 March 2025).

42 James MacCarthy, Jessica Richter, Sasha Tyukavina and Nancy Harris, 'The latest data confirms: forest fires are getting worse', *World Resources Institute*, 21 July 2025, available at <https://www.wri.org/insights/global-trends-forest-fires> (accessed 18 March 2025).

43 Lesley Hughes, James Rayner, Andrew Dean and Ketan Tidswell, *Dangerously overlooked: why we need to talk about methane* (Sydney: Climate Council, 2024),

- Should the global temperature rise 2 or even 3 degrees by the end of the century, if not before, the IPCC notes, this would lead to the complete melting of ice cover in West Antarctica and Greenland, causing *ocean levels to rise* by several metres. According to the UN and World Meteorological Society, if the oceans had not absorbed around 90 per cent of heat generated since the start of industrialisation, most Earthly life would probably have vanished by now. Certainly, marine heating and marine heatwaves have contributed to the destruction of some of the Earth's most precious ecosystems, not least the Great Barrier Reef.[44]
- Other feedback loops include *ice-albedo* feedback (exposure of darker ocean or land surfaces, which absorb more heat, leading to further warming and more ice melt); *water vapor* feedback where warmer temperatures increase evaporation, adding more water vapour to the atmosphere, which in turn traps heat and intensifies warming; *forest dieback* where rising temperatures and droughts weaken forests, reducing their ability to absorb CO_2, which increases atmospheric carbon levels; ocean *circulation changes* where warming disrupts ocean currents, affecting heat distribution and potentially weakening the ability of oceans to absorb CO_2; and *soil carbon release* where higher temperatures cause soil to release stored carbon, adding more CO_2 to the atmosphere and intensifying warming.[45]
- The *impacts of climate change are much more pronounced and widespread* than anticipated, already resulting in water shortages for nearly half of the world's population as well as diminished levels of crop production, threatening widespread food shortages. (According to a 2024 World Bank Group report,

---

available at <https://www.climatecouncil.org.au/resources/dangerously-overlooked-why-we-need-to-talk-about-methane/> (accessed 18 March 2025).

44 United Nations Regional Information Centre for Western Europe (UNRIC), 'Global warming: 90% of emissions heat absorbed by the ocean', 25 September 2023, available at <https://unric.org/en/global-warming-90-of-emissions-heat-absorbed-by-the-ocean/> (accessed 18 March 2025). See also: Great Barrier Reef Foundation (GBRF), 'Climate change is the single biggest threat facing the Reef', 2025, available at <https://www.barrierreef.org/the-reef/threats/climate-change> (accessed 18 March 2025).

45 William J. Ripple, Christopher Wolf, Timothy M. Lenton, Jay W. Gregg, Susan M. Natali, Philip B. Duffy, Johan Rockström and Hans Joachim Schellnhuber, 'Many risky feedback loops amplify the need for climate action', Alliance of World Scientists, 2023, available at <https://scientistswarning.forestry.oregonstate.edu/climate-feedback-loops-project> (accessed 18 March 2025). See also: Earthhow, '15 climate feedback loops and examples', 2023, available at <https://earthhow.com/climate-feedback-loops/> (accessed 18 March 2025).

the most food insecure populations due to crop failures caused by droughts and heatwaves are in Sub-Saharan Africa, South Asia, and Southeast Asia).[46]

- More *intense flooding* is already impacting a quarter of the world's population, and increasing heatwaves will displace millions of people, placing a heavy burden on poorer and recipient nations.[47] (This, as we have seen in the rise of ethnonationalism in Europe, has resulted in significant tensions both within and between nation states. 'Fortress' states may well tighten existing exclusion policies in order to placate angry electorates and to 'protect' economies).[48]

Recent monitoring efforts support the IPCC's findings, revealing that our planetary 'vital signs' are flashing red. According to *the 2025 state of the climate report: a planet on the brink*, of the thirty-four globally tracked indicators – including atmospheric CO_2 concentrations, ocean heat content, ice-sheet mass, tree-cover loss and extreme weather event frequency – 22 are now at record levels.[49] As the report states, unambiguously: 'we are hurtling toward climate chaos. The planet's vital signs are flashing red. The consequences of human-driven alterations of the climate are no longer future threats but are here now. This unfolding emergency stems from failed foresight, political inaction, unsustainable economic systems, and misinformation. Almost every corner of the biosphere is reeling from intensifying heat, storms, floods, droughts, or fires. The window to prevent the worst outcomes is rapidly closing'.[50] Importantly, the report further notes that global warming is not simply continuing, but appears to be accelerating, driven by

46 World Bank Group (WBG), 'What you need to know about food security and climate change', 17 October 2022, available at <https://www.worldbank.org/en/news/feature/2022/10/17/what-you-need-to-know-about-food-security-and-climate-change> (accessed 18 March 2025).

47 United Nations Office for Disaster Risk Reduction (UNDRR), *GAR 2025 hazard explorations: floods. global assessment report on disaster risk reduction* (2025), available at <https://www.undrr.org/gar/gar2025/hazard-exploration/floods> (accessed 18 March 2025).

48 Karl Mathiesen, Kalina Oroschakoff, Giorgio Coi and Aitor Busquets Guàrdia, 'Droughts, fires and floods: how climate change will impact Europe', *POLITICO Europe*, 2021, available at <https://www.politico.eu/article/how-climate-change-will-widen-european-divide-road-to-cop26/> (accessed 18 March 2025). Giovanni Bettini argued over a decade ago that 'climate barbarians' would face 'climate borders' presided over by 'fortress states' – a prediction that is being played out through various national immigration policies. Giovanni Bettini, 'Climate barbarians at the gate? the case for a cosmopolitan climate politics', *Geopolitics* 18/3 (2013), n.p., available at <https://doi.org/10.1080/14650045.2012.743905> (accessed 18 March 2025).

49 Ripple et al., 'Many risky feedback loops'.

50 Ripple et al., 'Many risky feedback loops'.

multiple compounding feedbacks including reduced aerosol cooling, increasing methane emissions, weakened land carbon sinks, and strong cloud responses.

More than half of the monitored variables – including sea-level rise, ocean acidity, and wildfire intensity – are now moving in directions that suggest irreversible damage to key planetary subsystems. Taken together, these indicators reveal that the Earth's system is moving beyond the relatively stable climatic conditions that have sustained human civilisations throughout the Holocene, signalling a 'phase shift' in the Earth's climate dynamics, in which the interplay of feedback loops – melting permafrost, deforestation, loss of sea ice, and ocean stratification – creates conditions for abrupt and self-reinforcing change.[51] The report frames this not only as an environmental catastrophe but as a civilisational one: a warning that humanity is rapidly exhausting the margin for error in a highly interconnected biosphere.

In outlining the need for urgent action on emission reductions, the IPCC notes that it *may already be too late* for many people and communities living near the equator, coastal regions, and low-lying islands. Depending on temperature rises, the climate catastrophe will render many places uninhabitable.[52] A report compiled by Chinese, US and European researchers predicts that by 2070, less than 50 years on current projections, up to 3 billion people could be living in regions exposed to Sahara-type temperatures.[53]

The overall picture presented by the IPCC has been confirmed by many other climate-watch organisations like the World Meteorological Organisation, which in its *State of the Global Climate Report 2023* noted that 2023 was the hottest year ever recorded, while in August 2024 the Copernicus Institute's Climate Change Service issued its latest *Climate Bulletin* finding that July 2024 was, 'the second-warmest July globally in the data record', averaging 1.49°C 'above the 1991–2020 average for July'.[54] This broke a 13-month consecutive streak of record-breaking temperatures. Meanwhile, sea surface temperatures for July 2024 were also the second highest on record, with parts of Europe both wetter than

51 Ripple et al., 'Many risky feedback loops'.

52 Josh Musto, 'Earth is becoming "increasingly uninhabitable," scientists warn', *The Independent*, 29 October 2024, available at <https://www.independent.co.uk/climate-change/news/earth-temperature-climate-uninhabitable-science-b2637796.html> (accessed 18 March 2025).

53 Chi Xu, Timothy A. Kohler, Timothy M. Lenton, Jens-Christian Svenning and Marten Scheffer, 'Future of the human climate niche', *Proceedings of the National Academy of Sciences* 117/21 (2020), 11350–55, available at <https://www.pnas.org/doi/10.1073/pnas.1910114117> (accessed 18 March 2025).

54 Copernicus Climate Change Service (C3S), *Climate bulletin*, 8 August 2024, available at <https://climate.copernicus.eu/climate-bulletins> (accessed 18 March 2025).

average or in prolonged drought, especially in southern and eastern Europe. Around the same time, a 15-month consecutive series of sea temperature records was broken, which resulted in mass coral bleaching around the world. Like the IPCC, the Copernicus Institute noted significant seasonal reductions in sea ice levels in both the Arctic and Antarctic.[55]

The evidence of rising temperatures is truly alarming. Drawing on consolidated datasets from a variety of sources, the World Metrological Organisation noted in early 2025, that all ten years from 2015 to 2024 were the warmest on record, and 2024 was the first full year to exceed 1.5°C, the symbolic Paris Agreement threshold.[56] These numbers, however precise, represent far more than statistical trends. They mark the boundaries of habitability, the thresholds beyond which familiar forms of life and governance begin to fail.

The response to this concerning scenario has been disappointing. The Copernicus Institute observes that governments across the world have generally failed to meet agreements set out in various COP forums. Despite some major successes like record renewable energy installations, overall progress in reducing emissions is slow and patchy. National Determined Contributions, which are the emission reduction and adaptation pledges made by countries under the Paris Agreement, are not ambitious enough, new fossil fuel projects are still being approved, governments continue to grant subsidies, and banks (as we shall see) continue to supply investment funding.

The net result is that emissions are still rising, albeit at slower rate. According to the Mauna Loa Observatory in Hawaii (which at the time of writing, the Trump administration plans to defund) the CO2 level in the atmosphere on 11 September 2025 was 424.61 ppm, a slightly less than one per cent-plus increase on the previous year which was already well above the 350-ppm safety threshold. This is roughly 50 per cent above pre-industrial levels, with the rate of increase 200 times faster than natural increases following the last ice age.[57] The problem with calculating carbon emissions levels, however, is that deforestation and other human-related activities, along with growing numbers of mega fires across the

55 C3S, *Climate bulletin*, n.p.

56 World Meteorological Organization (WMO), 'WMO confirms 2024 as warmest year on record at about 1.55°c above pre-industrial level', 10 January 2025, available at <https://wmo.int/news/media-centre/wmo-confirms-2024-warmest-year-record-about-155degc-above-pre-industrial-level> (accessed 18 March 2025).

57 Statista, 'Atmospheric CO_2 ppm by year 1959–2024', 7 August 2025, available at <https://www.statista.com/statistics/1091926/atmospheric-concentration-of-co2-historic/> (accessed 18 October 2025).

globe, mean it is difficult to accurately predict future levels of CO2 atmospheric concentrations.

If these sorts of observations seem repetitive, it is because they are: thousands of scientists have been saying the same and similar things for many decades.[58]

Joëlle Gergis is only too aware of what lies ahead. In a confronting 2024 article in the online journal, *The Conversation* titled 'An intergenerational crime against humanity: what will it take for political leaders to start taking climate change seriously?' Gergis states that, 'right now many nations' pledges have no finance, weak implementation or limited political ambition, so are effectively empty promises – there is a 90 per cent chance that we are still on track for 2.4°C of global warming under this best-case scenario, which will lock in centuries of irreversible changes to the climate system'.[59] Gergis notes, in another article, that, 'the dilemma we face is simple: genuinely respond to the scientific reality of our predicament or endure the consequences of this profound moral failure for centuries to come'[60].

The threat to Australia

For Australia, the prospect of escalating extreme weather events is clear, says Gergis: '... the geographical characteristics of Australia', she states, 'mean that the continent is warming faster than the global average – temperatures have risen by 1.5°C since 1910... As global warming continues, Australia's climate is fast becoming more extreme and unpredictable, edging us closer towards breaching thresholds that will make it very difficult, if not impossible, to adapt to... We see the continued degradation of our natural environment and the weakening

---

58 Alice Bell, 'Sixty years of climate change warnings: the signs that were missed (and ignored)', *The Guardian*, 5 July 2021, available at <https://www.theguardian.com/science/2021/jul/05/sixty-years-of-climate-change-warnings-the-signs-that-were-missed-and-ignored> (accessed 18 March 2025).

59 See also IPCC, *Climate change 2023*, which states that: 'The rise in weather and climate extremes has led to some irreversible impacts as natural and human systems are pushed beyond their ability to adapt'. NASA notes: 'The effects of human-caused global warming are happening now, are irreversible for people alive today, and will worsen as long as humans add greenhouse gases to the atmosphere'. NASA Science (NASA), 'The effects of climate change', [no date], available at <https://science.nasa.gov/climate-change/effects/> (accessed 18 March 2025).

60 Joëlle Gergis, 'Net zero distracts from scientific imperative to eliminate fossil fuels', *The Guardian* (7 November 2025), <https://www.theguardian.com/commentisfree/2025/nov/07/net-zero-distracts-from-scientific-imperative-to-eliminate-fossil-fuels-cop30>, accessed 7 November 2025.

of social resilience that will eventually lead to the permanent displacement of people from their homes and ongoing impacts on our economy'.[61]

In 2025 alone, Australia has experienced several major climate-related disasters in the form of 'unprecedented' floods in inland Queensland and the central coast, and Cyclone Alfred in March that impacted southeast Queensland and northern NSW, causing damage estimated at $1.2 billion.[62] The flooding that occurred on Australia's eastern seaboard in 2022 is estimated by professional services firm Deloitte at around $7.7 billion.[63] Faced with all this, it is hardly surprising that the prospect of ecological collapse looms so large in the minds of scientists, activists, responders, insurance companies and others.

In its 2023 report titled, *Mission Zero: How today's climate choices will reshape Australia*,[64] the Canberra-based Climate Council observed that, 'climate change is assaulting many aspects of the great Australian way of life': from the increasing dread of summer heat, fires, floods and storms, to the destruction of reef ecologies and ancient Gondwanan forests, our collective wellbeing is under severe strain. Just about everything and everyone is at risk. This includes our mental health, economic security, and social stability. Even many of Australia's iconic beaches are predicted to disappear due to sea level rises, and some cultural events like music festivals, once a regular feature on the cultural landscape, have already disappeared, partly as a result of extreme weather events.[65] The impacts of a changing climate will, according to the Council's report, be experienced differently, especially impacting our most exposed populations – people with disabilities, Indigenous people, the poor and marginalised.

This grim outlook is supported in a report by the Australia Academy of Sciences (AAS) which among other things, notes that extreme weather events

---

61 Gergis, *'Highway to hell'*.

62 Insurance Council of Australia (ICA), 'Catastrophe report: cyclone Alfred – March 2025', 2025, available at <https://insurancecouncil.com.au/reports/cyclone-alfred-catastroph e-report-march-2025> (accessed 30 March 2025).

63 Deloitte Access Economics, *South East Queensland rainfall and flooding event – impact assessment*, 2022, commissioned by the Queensland Reconstruction Authority, available at <https://statements.qld.gov.au/statements/95831> (accessed 18 March 2025).

64 Simon Bradshaw, Lesley Hughes, David Karoly, Joëlle Gergis, Annika Dean, Dinah Arndt, Jennifer Rayner and Jolee Wakefield, *Mission zero: how today's climate choices will reshape Australia* (Climate Council of Australia, 20 September 2023), available at <https://climatecouncil.org.au/resources/missionzero/> (accessed 18 March 2025).

65 Timothy Mortlock, 'How rising sea levels will affect our coastal cities and towns', *Australian Geographic*, 24 January 2024, available at <https://www.australiangeographic.com.au/news/2024/01/how-rising-sea-levels-will-affect-our-coastal-cities-and-towns/> (accessed 18 March 2025).

will become more frequent as the climate warms. Alarmingly, the Academy notes that, 'the total emission reductions currently pledged by the Australian and international governments through the ... Paris Agreement ... even if implemented on time, will translate as average global surface temperatures of 3°C or more above the pre-industrial period by 2100'.[66]

Australian marine biologist Ove Hoegh-Guldberg and climate scientist Lesley Hughes invite us to, 'imagine, for a moment, a different kind of Australia. One where bushfires on the catastrophic scale of Black Summer happen almost every year. One where 50°C days in Sydney and Melbourne are common. Where storms and flooding have violently reshaped our coastlines, and unique ecosystems have been damaged beyond recognition – including the Great Barrier Reef, which no longer exists. Frighteningly, this is not an imaginary future dystopia. It's a scientific projection of Australia under 3°C of global warming'.[67]

At 3°C of global heating, the AAS observes (a figure slightly higher than IPCC assessment), 'many of Australia's ecological systems would be unrecognisable', changing its landscape forever. With regard to agriculture, forestry and fisheries, the report notes that droughts will intensify in southern eastern and southwestern Australia, meaning less water availability for crops, while heat stress will impact livestock health, grazing and crop yields, significantly affecting the viability of many farming businesses – and, of course, impacting food supply. More extreme heat will also threaten Australian forests, leading to more fires, thereby adding to atmospheric pollution, while fishery stocks will be significantly lower as a result of acidification and ocean warming. Across all agricultural industries, the AAS states, a 3-degree rise in temperature would greatly impact production levels and therefore increase food insecurity.[68]

Given the concentration of Australia's population in cities and towns, mainly on the eastern seaboard, more people will be impacted adversely by climate-related events, with the consequence of diminished property values, damaged infrastructure, overburdened energy supply systems, excessive demands on health services,

---

66 Australian Academy of Science (AAS), *The risks to Australia of a 3°c warmer world* (Canberra: Australian Academy of Science, 2021), available at <https://www.science.org.au/supporting-science/science-policy-and-analysis/reports-and-publications/risks-australia-three-degrees-c-warmer-world> (accessed 18 March 2025).

67 Ove Hoegh-Guldberg and Lesley Hughes, 'Seriously ugly: here's how Australia will look if the world heats by 3°c this century', *The Conversation*, 31 March 2021, available at <https://theconversation.com/seriously-ugly-heres-how-australia-will-look-if-the-world-heats-by-3-c-this-century-157875> (accessed 18 March 2025).

68 AAS, *The risks to Australia of a 3°c warmer world*, n.p.

unaffordable insurance premiums, business insecurity and disruption to social and community life.[69]

(We know from bitter personal experience that increased rainfall in northern NSW – one report pouts the likely increase by the end of the century up to 70 per cent – would mean more of the catastrophic flooding we both experienced in 2017 and 2022).[70]

In short, Australia will be a very different place to what it is today. The nation seems unwilling or unable to confront the very real possibility that parts of the country may be uninhabitable, and that every aspect of life will be altered. Given the impact of rising oceans on coastal communities and the fact that many towns and cities have rivers running through them, means that the financial costs of adaptation will be huge, with many councils in Australia already struggling.[71]

Such problems need to be considered in the context of many parts of the country already under great financial strain. Higher food prices and other 'cost of living' expenses resulting from climate impacts will therefore place more pressure on already heavily indebted households, many of which are already facing high rental costs and huge mortgage repayments.[72] Australia is, according to the Reserve Bank of Australia, already one of the most indebted societies in the world, mainly due to housing debt.[73] Moreover, when it comes to Australia's major cities, the median price-to-income ratios is among the highest in the world, making house-buying out of reach for the majority of middle-income earners. Additionally, there is also growing wealth and income inequality in Australia – in 2025 reaching a 20-year high – along with over three and a half million people

---

[69] Commonwealth Scientific and Industrial Research Organisation (CSIRO) and Bureau of Meteorology (BOM), *State of the climate 2024* (Commonwealth of Australia, 2024), available at <https://www.csiro.au/en/research/environmental-impacts/climate-change/State-of-the-Climate/Australias-Changing-Climate> (accessed 18 March 2025).

[70] NSW Department of Planning and Environment (DPE), *NARCliM climate change projections for New South Wales* (2021), available at <https://www.climatechange.environment.nsw.gov.au/climate-projections> (accessed 17 March 2025).

[71] The Australia Institute, 'Local governments face soaring cost of climate change', 10 September 2025, available at <https://australiainstitute.org.au/post/local-government s-face-soaring-cost-of-climate-change/> (accessed 28 September 2025).

[72] The Australia Institute, 'Climate crisis escalates cost-of-living pressures', media release, 17 February 2025, available at <https://australiainstitute.org.au/post/climate-crisis-escalate s-cost-of-living-pressures/> (accessed 18 March 2025).

[73] Jonathan Kearns, Mike Major and David Norman, 'Why is Australian household debt high relative to history and other countries?', *Research Discussion Paper* 2020-05, Reserve Bank of Australia (RBA), 2020, available at <https://www.rba.gov.au/publications/rdp/2020/2020-05/why-is-australian-household-debt-high-relative-to-history-and-other-countries.html> (accessed 18 March 2025).

in poverty.[74] This means that many sections of Australian society find themselves in precarious economic circumstances which are likely to worsen under deteriorating climatic conditions.

The terrifying reality of what the climate catastrophe means for Australians has been outlined by the Australian Climate Service which, on 15 September 2025, published the first government-commissioned National Climate Risk Assessment.[75] It makes for harrowing reading, confirming most of the above findings. A 2.7-degree hothouse Earth will drastically alter Australians' way of life. Reaching tipping points will abruptly lead to uneven yet catastrophic changes. 'It is likely', the report states, 'we will experience more compounding, cascading and concurrent hazards'. Among many consequences already mentioned, the report says that by 2050 – less than 25 years from now – 1.5 million people living in coastal areas will experience increased flood risks. Urban, remote and regional areas and the poor, marginalised and Indigenous people are most at risk of extreme weather events, the report says.

In addition to reconfiguring Australian geographies, environmental damage will massively increase government spending and disrupt social cohesion (a point mentioned but not elaborated in the report). However, the situation might be even worse than expected as the change in distribution, timing and severity of extreme weather events means that historical observations on their own are not likely to be a good indicator of future risk.[76] In effect, this means that the cascading effects of environmental destruction remain unknown. The breaching of tipping points and feedback loop effects make accurate forecasts difficult. Yet so concerned is the Australian Federal Government about what the climate disaster will bring that it has blocked the release of another report on climate security implications by the Office of National Intelligence (ONI) which, according to the few Members of Parliament who have viewed it, say its contents are 'terrifying' and 'horrible'.[77] Commissioned by the Albanese administration, the report details security threats posed by escalating climate destruction. Unlike the disturbing findings of the Climate Risk Assessment, the ONI's report was clearly deemed a bridge too far for

---

[74] University of Melbourne, 'HILDA survey shows inequality rises to a high', *Newsroom*, 2025, available at <https://www.unimelb.edu.au/newsroom/news/2025/march/hilda-shows-inequality-rises-to-a-high> (accessed 8 April 2025).

[75] Australian Climate Service, 'National Climate Risk Assessment'.

[76] Australian Climate Service, 'National Climate Risk Assessment'.

[77] Karen Barlow, 'Exclusive: government refuses to release climate security report', *The Saturday Paper*, 20 September 2025, available at <https://www.thesaturdaypaper.com.au/news/politics/2025/09/20/exclusive-government-refuses-release-climate-security-report> (accessed 12 October 2025).

public consumption – a decision that ultimately places Australian communities in even graver danger.[78]

Such developments only serve to deepen public anxieties. People and communities will have to face up to what Earth scientists have long known – namely, that systems breakdown will present the greatest challenges ever faced by humankind.

Conclusion

What emerges from this chapter is a portrait not just of environmental decline, but of societies struggling to metabolise its own knowledge. The scientific evidence is unambiguous and overwhelming, yet our collective response remains paralysed by denial, distraction, and delay. The environmental disaster has moved beyond a question of awareness; it is now a crisis of meaning and imagination. To live in this time is to know more than any previous generation about the fragility of Earth's systems, while being demonstrably unable or unwilling to change course.

For scientists such as those mentioned here, the trauma of engaging daily with climate data has become both personal and professional. Their distress exemplifies a deeper pattern of collective wounding – a form of cultural and ecological trauma arising from prolonged exposure to loss, uncertainty, and inaction. These traumatic responses reveal the extent to which climate disruption has become an existential conundrum as well as a profound crisis of perception and purpose. To know the scale of planetary decline, and yet remain constrained within systems incapable of responding adequately, is itself a destabilising and disorienting experience.

For us, the notion of a 'Great Unravelling' thus captures more than the collapse of ecosystems. It speaks to the erosion of assumptions that once sustained modern industrial societies: faith in technological progress, economic growth, and human exceptionalism. This brings us to a pressing question: in light of the gravity of our global predicament, how does the extractivist system of consumer capitalism continue to persist? Understanding how this cannibalistic system endures – despite overwhelming evidence of its self-destructive impulses – becomes essential to any honest reckoning with our future. It is to this question, and to the political and economic logics that sustain the Great Unravelling, that the next chapter now turns.

---

78 Barlow, 'Exclusive: government refuses to release climate security report', n.p.

Business as usual? Hypernormalisation and the climate disaster

The most likely manner in which mankind is influencing the global climate is through carbon dioxide release from the burning of fossil fuels.

—EXXON RESEARCH & ENGINEERING COMPANY, 1982

Business as usual has brought us to the edge of extinction.

—RUPERT READ

Introduction

When catastrophic floods struck the Northern Rivers in 2022, Lismore residents like Catherine Barker ('Cat') confronted not only physical devastation but a profound sense of loss and disorientation. The disaster shattered assumptions of normality, even though public discourse remained curiously muted. Many politicians hesitated to link the floods explicitly to climate disruption, while much of the media framed them as 'natural disasters' rather than symptoms of systemic disruption. Cat's question – '*Why are we pretending the situation is normal?*' – captures the unease of a society caught between knowing and denial.

This chapter examines that tension. We argue that the collective performance of 'business more or less as usual' is not accidental but the product of hypernormalisation: a condition in which everyone recognises the system's demonstrable failure, yet continues to act as if it were functioning normally. Based on the understanding that domination depends to a large extent on consent, we explore how political, economic, and cultural forces sustain the illusion of normality even as the world unravels.

By unmasking these entanglements, we show how denial and obfuscation become institutionalised and profitable, transforming catastrophe into profit-generating business opportunities. Yet as fires, floods, and heatwaves multiply, the cracks in the façade have begun to widen, offering us a glimpse of the exhaustion and moral bankruptcy of a system that devours the future in order to preserve the present.

Cat's story

Cat lives on one of the highest points in Murwillumbah, a town in northern NSW that has seen its fair share of flooding over the years. She feels safer there, much safer at least than where she lived before – in Lismore, a small regional town known to be the most floodable town in the country, that in 2022 experienced one of the most devastating floods ever seen in Australia.

Cat's not someone who is easily shocked, but what happened nearly four years ago in Lismore still haunts her. After many years in the entertainment industry, Cat moved to the Northern Rivers area and felt settled there. She '*loved the town*' – a place '*full of interesting, creative people*'. '*There was a real sense of community there*', Cat said.

When she woke on the morning of 28 February 2022 the east side of town where she lived had not yet been impacted by the floods. She was picking up stories of numerous people in the south and north of the town who had to swim out of their homes.

Gradually though, the full enormity of the unfolding disaster began to dawn on her. Cat also knew that her own home could be impacted.

> *When I realized the extent of what was going on, I was shell-shocked. No-one predicted this, not BOM or the government. I was surprised because Lismore knows floods, people know what to do, how to prepare... But no-one expected this... We were all working off the last floods, but this one was so much worse, really shocking and terrifying.*

'*All I could really think about*', Cat said, as the flood waters began to recede,

> *was how to help my clients and friends (Cat works in disability services). I went back a day or so later and was on the frontline. I tried to find clients in evacuation centers. We had over 300 clients. So many were affected. They were highly vulnerable people, with multiple needs, many requiring urgent medication. We'd lost four venues, and our server. We were all trying to find solutions for people who had nowhere to stay'.*

> *'Looking back, the worst thing was that I was struggling at the time with my mum dying. It was a really difficult time. The flooding made things so much worse.*

Cat had to make some tough decisions:

> *It was not long after the floods that I decided to head to Sydney to be with my mum... I also became aware of how little people in Sydney understood the situation in Lismore. They really didn't know what the flood meant and how it impacted the community. I think that's when I decided to make a documentary called* When the river rose. *(Cat had worked in TV production for several years). It recounts people's stories of the floods and the aftermath. People who've watched it have been very moved. And of course, when it was shown in Lismore we had a packed house. Making the film was really challenging but I felt I had to get those stories across to people who didn't really understand how bad it was.*

Cat returned to Lismore after her mum passed away, but things had changed too much for her to stay.

> *I was affected by being there, especially when the rain came. It was very intense. I decided to leave, so I bought a house on highest point in Murwillumbah. When cyclone Alfred happened in March this year, I found myself worrying about Lismore. I knew people in there were on edge. I was too. I was so stressed. It happens when there are large spates of rain...*

'*Looking back*', Cat continued

> *all the loss and grief was on top of the grief for my mum. There was so much grief around. It was a very sad situation, so many places and people have been impacted... I've learnt to push aside the trauma. I speak of it from the periphery, but it sits in the nervous system. It's absolutely terrible to see so much suffering ... It stays with you... I don't get why we're not talking about these sorts of events more openly. Why are we pretending the situation is normal?*

Cat's bewilderment at the calculated refusal to name the catastrophe foreshadows a larger pattern explored in this chapter – namely, the collective performance of normality that persists even as the world burns.

Cat's experience of the 2022 Lismore floods resonates with thousands of other Australians in similar circumstances. Yet oddly, even in the midst of such catastrophic events, it seems that few people are prepared to talk openly and honestly about the full grim reality of 'anthropogenic climate change' – and the likelihood that things are only going to get worse. The media self-censor, and politicians avoid 'politicising' a 'tragedy'.[1] Still, there is an unspoken sense that

---

[1] For discussion on such matters, see: Paul Blamey, 'Climate change: media, politicians need to change the rhetoric', *Crikey* (8 April 2024), <https://www.crikey.com.au/2024/04/08/climate-change-rhetoric-media-politics/>, accessed 5 April 2025. Australian politicians have also refused or been reluctant to speak about extreme weather events and 'climate change': Bernard Keane, 'Election '25: two campaigns and absolutely no winners', Crikey

these 'unprecedented', 'record breaking', 'apocalyptic' and 'biblical' events are part of an unfolding and deeply troubling new reality. Places ravaged by extreme weather events – and there are now many of them – become ghoulish symbols of an emergent dark future.

Lismore has not – perhaps never truly will – recover from the events of 2022.[2] Cat misses the place. But now it is all about the memories. Many shops and houses remain boarded up. Parts of the town have a ghostly feel. Traumatised onlookers like Cat are asking why the causes of such disasters are not being addressed in the way they should be. She's right to wonder.

Feeding a system that eats itself

The rest of this chapter addresses Cat's concern: namely, how, despite its obvious and life-threatening flaws, does extractivist capitalism persist? Let us go further: given what we already know about the origins and consequences of our ecological predicament, why are fossil fuel companies permitted to continue selling their lethal wares?

While there is much talk about 'transition' and policies designed to achieve 'net zero', and despite Australia's claim to being a colossus of 'clean energy' – and there is some truth to this (as there is in China) – the energy system as it exists, and indeed the way of life associated with it, relies on the very modalities of production and consumption that have led to this point. More on this later. At this juncture, we examine the ideological frameworks that continue to legitimise and perpetuate the prevailing system.

There are several things to consider here. The first is the broad commitment of powerful economic and political interests to the continuation of the present system. This is achieved not through the actions of a secretive cabal, but rather through common interest and adherence to a particular way of doing things. Integral to this is an unwavering devotion to the basic tenets of capitalism: private ownership of the means of production, capital accumulation, compound economic growth, market competition, high levels of productivity, and mass consumption. Energy supply – clean and dirty – is the engine of this system. It

---

(2 May 2025), <https://www.crikey.com.au/2025/05/02/2025-federal-election-anthony-albanese-peter-dutton-campaigns/>, accessed 5 April 2025.

2 For an outline of the enduring problems this country town, see: Lisa White and Ellie Rennie, 'Have we learned the lessons from the 2022 Lismore flood disaster?', *ABC News* (27 February 2023), <https://www.abc.net.au/news/2023-02-27/have-lessons-been-learned-from-2022-lismore-flood-disaster/101863278>, accessed 5 April 2025.

is what literally keeps it going. Governments continue to support this industry through various financial and other means. Indeed, as we outline below, there exists an unhealthy symbiosis between these entities, evidenced when governments offer subsides and companies reciprocate with political donations.

Second, as noted earlier, the capitalist system – and its latest iteration, neoliberal capitalism – has been promoted as the only credible and workable arrangement available to humankind, just as liberal democracy was once touted as the 'end of history'.[3] Third, the capitalist system has demonstrated a remarkable chameleon-like capacity to change its colours, to absorb difference and dissent, to reinvent itself, and to present itself as benign, natural and effective – despite all evidence to the contrary. Fourth, and relatedly, alternatives to the capitalist system – or fundamentally different socio-economic arrangements – are considered either unworkable and/or dangerous, presenting a threat to social order.[4] Fifth, to manage the system and to deflect dissenting voices, the system remains reliant on manufactured consent. After all, no system can function easily with ongoing, pervasive opposition (some do, of course). Methods of achieving order and conformity have to be implemented. One way of doing this is to ensure that social, economic, and political discourses occur within accepted parameters which, in the case of neoliberal capitalism, means adherence to the language, values and practices of the normative order. For example, few if any politicians in Australia – or in any other western state for that matter – would dare to openly question the orthodoxy of economic growth – to do so would end their careers.

When it comes to one of the most harmful by-products of the present system – the climate catastrophe – the discursive framing takes an interesting turn. It involves diminishing the 'problem' and insisting that it can be fixed through capitalist ingenuity, along with various market manipulations. Despite the gravity of the threats we face, there is little or no acknowledgement among economic and political elites that it may already be too late to prevent a hothouse Earth.[5] To

---

3 See Francis Fukuyama, *The end of history and the last man* (New York: Free Press, 1992).

4 Sociologist David Lane notes that this has been a common refrain among advocates of free market capitalism, expressed most forcefully in Margaret Thatcher's famous dictum that 'there is no alternative'. David Lane, *Global neoliberal capitalism and the alternatives: from social democracy to state capitalisms* (Bristol: Policy Press, 2023).

5 In 2021 former CEO of the Australian Institute of Company Directors, Ian Dunlop, noted that, 'it is probably too late to avoid major disruption, but we must act urgently to prevent the worst'. He calls for economies to be placed on a 'war footing'. Since 2021, emissions are still rising. Ian Dunlop, 'Climate change and politics as usual – a government missing in action', *Ian Dunlop website* (7 September 2021), <https://itdunlop.com/climate-change-politics-as-usual-a-government-missing-in-action/>, accessed 5 April 2025.

admit this would also be politically fatal. Instead, the narrative that swirls around policy circles relies heavily on what sociologists Arthur Mol and Gert Spaargaren refer to as ecological modernisation theory – the belief that environmental degradation can be decoupled from economic growth through technological innovation, efficiency, and market-based regulation.[6] According to this view, capitalism is not considered the cause of the ecological crisis but rather its most effective remedy – an odd double act of perpetrator and saviour. This routine is also played out in the green growth paradigm, popularised by institutions such as the OECD and the World Bank,[7] which claim that economies can continue expanding while reducing emissions through investment in renewable energy, carbon markets, and 'sustainable' infrastructure. As critical scholars like John Bellamy Foster, Jason Hickel, and Giorgos Kallis have argued,[8] these contradictory approaches serve as powerful ideological instruments: they promise transformation without disruption, growth without guilt. In so doing, they depoliticise the climate predicament, redirecting attention toward technocratic solutions while leaving the underlying logic of accumulation intact. The result is a form of adaptive capitalism that converts existential threat into economic opportunity – where 'net zero', 'clean energy transitions', and 'carbon offsetting' become the latest expressions of business as usual. It is the arena in which ecological life can be purchased and traded through 'biodiversity offsets', 'habitat banking', 'pollination credits' or 'blue bonds'. The 'clean energy revolution', we are told, will save us. All that is required is the political will.[9]

A more recent expression of this adaptive logic can be seen in the rise of artificial intelligence. Far from being an immaterial innovation, AI relies on vast physical infrastructures: data centres, energy-hungry servers, cooling systems, rare-earth minerals, and global supply chains that mirror the extractive dynamics of the fossil-fuel era. Training large language models consumes significant amounts

---

6 Arthur P. J. Mol and Gert Spaargaren, 'Ecological modernisation theory in debate: a review', Environmental Politics 9/1 (2000), 17–49.

7 Organisation for Economic Co-operation and Development (OECD), *Towards green growth* (Paris: OECD, 2011); World Bank, *Inclusive green growth: the pathway to sustainable development* (Washington, DC: World Bank, 2012).

8 John Bellamy Foster, *The ecological rift: capitalism's war on the earth* (New York: Monthly Review Press, 2011); Giorgos Kallis, *Limits: why Malthus was wrong and why environmentalists should care* (Stanford: Stanford University Press, 2019); Jason Hickel, *Less is more: how degrowth will save the world* (London: William Heinemann, 2020).

9 This view was aired by a number of Australian commentators. Zudy Zel see this revolution as 'the cornerstone of economic growth'. Zoe Zel, 'The future of renewable energy: election promises for 2025', *Australia Unwrapped* (2 April 2025), <https://www.australiaunwrapped.com/the-future-of-renewable-energy-election-promises-for-2025/>, accessed 5 April 2025.

of electricity and water, intensifying pressures on already stressed ecosystems. These systems are owned and operated by a handful of technology giants, deepening corporate concentration and expanding economic inequality. While AI has genuine potential to support emissions reduction, disaster modelling and ecological monitoring, its benefits depend entirely on the context in which it is deployed. Without strict governance, transparency and public oversight, these innovations risk reinforcing the same extractive and growth-driven dynamics that underpin the climate crisis.

To bolster the normalisation agenda – albeit with some necessary technical adjustments – clever cultural messaging is required. This can take many forms – from outright propaganda to more subtle techniques of persuasion. Generally, proponents of the system insist on its basic integrity. Told repeatedly enough, this illusion tends to hold sway. The result is a shared fiction which insists that the dominant order, despite all its flaws, remains not only inevitable but desirable. The deeper the disruption becomes, the stronger the compulsion to perform the act of normalcy – a choreographed display of collective self-reassurance that allows daily life to proceed undisturbed by the enormity of what is unfolding.

Italian theoretician, Antonio Gramsci, explored this logic in his 1929 *Prison Notebooks*, arguing that hegemonic domination occurs through 'the active consent of the governed' who ultimately come to regard the dominant system as natural, normal, and inevitable'.[10] British documentary film maker, Adam Curtis, has taken this argument a step further. Following Soviet dissident Alexei Yurchak, Curtis evokes the notion of 'hypernormalisation' which, he points out, seeks to represent, 'total fakeness as normal' – mainly as a way of diverting attention from core issues and affirming the indispensability of the prevailing system.[11] In Yurchak's original formulation, hypernormalisation described the late Soviet Union's condition in which everyone knew the system was failing, yet continued to assert it was functioning normally, thereby maintaining a collective pretence of stability in the face of obvious decay.[12] In contemporary neoliberal societies, this same logic manifests as a perverse form of cognitive dissonance articulated through the widening gap between the reality of planetary breakdown and an insistence on the preservation of the status quo. Citizens, politicians, and corporations alike become participants in this act of self-delusion, performing

---

10 Antonio Gramsci, *Selections from the prison notebooks*, ed. and trans. Quintin Hoare and Geoffrey Nowell Smith (London: Lawrence & Wishart, 1971).

11 Adam Curtis (dir.), *HyperNormalisation* (BBC, film, 2016).

12 Alexei Yurchak, *Everything was forever, until it was no more: the last Soviet generation* (Princeton: Princeton University Press, 2005).

normalcy in order to maintain the current order of things. Hypernormalisation thus operates not merely through crude propaganda but via everyday routines, mediated narratives, and institutional inertia that render claims of systemic collapse as unthinkable and even treacherous.

Sometimes though, as Curtis observes, cracks appear in the system, briefly exposing its true nature. 'Whenever journalists point [the problems] out, Curtis argues, "everyone goes, 'wow, that's terrible!' – and then nothing happens, and the system remains the same"'.[13] Similarly, the 'problem' of 'climate change' is seen as temporary, fixable problem in an otherwise well-oiled machine. And, as in times of war, the 'problem' of 'climate change' can be turned into a profit-making opportunity as in creative schemes like carbon credits, emissions trading schemes and various related 'clean energy' initiatives that rarely challenge the underlying paradigm of growth and consumption.[14]

Such questionable practices aside, the illusion of normalcy can only hold for so long. Rising dissent and repeated climate disasters expose more and more cracks in the system. Existing tensions are exacerbated, and new fault lines appear. The zeitgeist becomes ominous, people begin to complain openly, with some taking to the streets. Others of course, draw the curtains and pretend that none of this is happening. History shows however that when normative illusions begin to melt away or shatter, systems can collapse – often spectacularly. What once appeared unassailable, can soon topple into oblivion. As the title of Yurchak's book suggests: *Everything Was Forever, Until It Was No More.*[15]

That said, unstable and comprised systems can and do endure – sometimes for long periods. Much depends on the power of illusions and how authority is exercised. Societies can also appear robust and organised, until structures begin to unravel. At times, the process of disintegration is hidden, covered over by myths and illusions. Our view is that Australia, like many other western nations, is in this space of deep uncertainty: caught in the fiction that 'climate change' is manageable while also sensing that everything has changed, and that nothing is certain any longer.

At first, this tension may be played out through subtle shifts in consciousness. For instance, since the collective trauma of the 2019–2020 Black Summer bush-fires, Australians' relationship with summer has been markedly unsettled. Once

13 Curtis, *HyperNormalisation*.

14 The climate disaster thus becomes monetised, often serving the interests of capital rather than the environment. See Brett Christophers, *The price is wrong: why capitalism won't save the planet* (London: Verso Books, 2022).

15 Yurchak, *Everything was forever.*

celebrated as the highlight of the year – a season of holidays, beaches, barbecues, and long, carefree days – it now carries an undercurrent of unease. The familiar anticipation has given way to apprehension, a quiet vigilance about what each summer might bring: bushfires, floods, cyclones, or droughts? Our sense of summer has shifted – subtly, yet unmistakably – from one of joy and abundance to an experience shadowed by threat and uncertainty.

Another example may help us comprehend the bigger, often-unacknowledged changes that course through our minds. One of the authors, Richard Hil, was a backseat passenger in a car with two friends who together had spent two enjoyable hours talking about what constituted 'truth'. As we trundled toward our destination, the conversation turned to the subject of the 'climate emergency'. The friends were asked how they felt about this, whether it had altered the way they thought about their place in the world. '*Oh yes, all the time*', one said with unexpected gusto, adding that, '*I think we're heading for total disaster and there's no way of getting out of it*', while the other joined in, saying, '*yes, that's more or less how I feel. I really don't hold out much hope. None in fact. We just have to live with this and see what happens. None of us know what's going to happen really, other than the world won't be like it is today. That's for sure*'. It turned out that both of these individuals had held these views for a number of years, but rarely aired them, choosing instead to let other aspects of their lives take precedence – even though unexpressed anxieties had lodged deep within. Following a short, contemplative silence, one of the friends continued: '*all that we have really, or what we should be investing in, I think, is love and belonging, and perhaps to work on our connections with others as this shit works its way through*'. A minute or so later, he added that, '*what I don't understand, what really gets me, is how the people most responsible for what's happening get away with it, how they live with themselves, how they continue to place us all in such great danger*'.

These are pressing matters, indeed – core to our present discussion. So, what actual practices sustain the current system going? Who benefits? And why? More specifically, what enables the driver of capitalism – the dirty energy industry – to continue along its destructive path? To answer these questions, we now examine how various institutional arrangements have over time have propped up the fossil fuel sector – and indeed supported an entire way of life based on the imperatives of economic growth and capital accumulation.

Propping up dirty energy

Nominally at least, it is the task of governments to manage the resources of the nation-state in ways that avoid harming its citizens, or indeed, other people around the world. Despite Australia's much-vaunted commitment to a 'clean energy future' and to becoming a 'renewable energy superpower',[16] its practices are glaringly contradictory. Successive Australian governments have continued to support fossil fuel extractivism. They have done so in the full knowledge of the climate threat. As eminent climate scientist and reviewer of numerous IPCC reports, Peter Caster and environmental journalist, Elizabeth Woodworth, point out in *Unprecedented Crime: Climate Science Denialism and Game Changers for Survival*, '... they [governments] simply cannot say they did not know ... national governments are undeniably culpable for their continued lack of action to bring about a rapid decline in global emissions'.[17]

Not only is Australia the world's third largest exporter of fossil fuels, after Russia and the US, it is also committed to the maintenance and expansion of extractivist projects into the future. To explain Australia's contradictory stance, it is necessary to understand the complex interconnections between governments and the energy sector which hinge on a range of financial and political interests.

Subsidies

Australian governments prop up fossil fuel companies through a range of lavish subsidies. In 2024, the progressive think-tank The Australian Institute, spelt out precisely how much state and federal governments handed to fossil fuel industry in the form of subsidies, totalling $14.5 billion in 2023–24 – well up on the $11.1 billion recorded in the previous financial year.[18] Over the longer term, these subsidies have totalled around $65 billion. As the Institute reports, this is sixteen times the amount given to Australia's Disaster Ready Fund. The Federal Government is the biggest provider, handing over a total of $54 billion.

---

16 Current Australian Federal Minister for Climate Change and Energy, Chris Bowen, told a meeting of State and Federal Climate and Energy Ministers in July 7, 2023 that, 'I'll tell you what the future of Australia is – a renewable energy superpower, not a nuclear back-water'. Chris Bowen, 'Energy and climate change ministerial council press conference', *Department of Climate Change, Energy, the Environment and Water* (7 July 2023), <https://minister.dcceew.gov.au/bowen/transcripts/energy-and-climate-change-ministerial-council-press-conference>, accessed 05 April 2025.

17 Peter D. Carter and Elizabeth Woodworth, *Unprecedented crime: climate science denial and game changers for survival* (Atlanta, GA: Clarity Press, 2017).

18 Richard Campbell, Evan Morison, Mark Ryan, Millie Saunders, My N. Le, Ashim Adhikari, Kade Scicluna, Ellen Simpson and Lisa Anderson, *Fossil fuel subsidies in Australia 2024* (13 May 2024, The Australia Institute), <https://australiainstitute.org.au/report/fossil-fuel-subsidies-in-australia-2024/>, accessed 6 April 2025.

According to finance academic Bernard Njindan Iyke, subsidies to fossil fuel producers vary from tax breaks and direct payments to grants and other payments which are considered vital stimulants to wider economic activity in areas like manufacturing, transport and power. In addition to direct subsidies, there are what Iyke refers to as 'implicit subsidies'.[19] These refer to the costs borne by climate-impacted communities in Australia and elsewhere. 'Most fossil fuel companies', says Iyke, 'don't have to pay a cent for the pollution their products cause – so in effect, they are being granted an indirect subsidy'. That's not all. Governments also bend over backwards to support the necessary infrastructure for fossil fuel production. This includes 'investment in transport networks, pipelines, oil refineries and port infrastructure'.[20]

In the wake of a 31 per cent hike in subsidies paid out in 2024, Ron Campbell and co-researchers at the Australian Institute conclude: 'The subsidies that Australian governments give to the fossil fuels industry come at the expense of action on climate change. Existing policies supposedly designed to tackle climate change – like carbon offset systems, land sector accounting, hydrogen, and carbon capture and storage – merely greenwash the continuing, increasing support that Australian governments give the fossil fuels industry'.[21]

The net effect of this is to ratchet up extreme weather risks. As Professor of Law at Notre Dame University Bruce Huber notes, government subsidies, 'undermine efforts to address climate change because they make fossil fuels cheaper than they would be otherwise', meaning that climate polluters think their investments in dirty energy are worth it.[22] Indeed, the profits made by these companies would suggest they have a point, even though the cost has to be borne by countless others. Campbell and colleagues argue that: 'Ending fossil fuel subsidies should be the starting point of any sincere climate policy. Cutting fossil fuel subsidies would not only help achieve genuine reductions in emissions but would save money that could be spent on public services'.[23]

---

19 Blessing N. Iyke, 'For decades, governments have subsidised fossil fuels. But why?', *The Conversation* (30 July 2024), <https://theconversation.com/for-decades-governments-have-subsidised-fossil-fuels-but-why-213467>, accessed 6 April 2025.

20 Iyke, 'For decades, governments have subsidised fossil fuels'.

21 Campbell et al., Fossil fuel subsidies in Australia 2024, 65.

22 Benedikt Huber, 'Countries spend huge sums on fossil fuel subsidies – why they're so hard to eliminate', *The Conversation* (13 November 2024), <https://theconversation.com/countries-spend-huge-sums-on-fossil-fuel-subsidies-why-theyre-so-hard-to-eliminate-239346>, accessed 6 April 2025.

23 Campbell et al., *Fossil fuel subsidies in Australia* 2024, 65.

Political donations

Another form of entanglement involving governments and polluting industries is the latter's questionable influence on policy making via donations. A key driver of this unholy alliance is the hefty amount of donations handed over to governments by polluting companies which, in the financial year ending in 2022, totalled $2 million to the ALP, Liberal and National parties.[24] With over a third of donations coming from unidentified sources, the actual figure is probably considerably higher. Companies like Woodside, the biggest donor, pursuant of a number of gas and oil projects, including the controversial Scarborough Gas project, donated $2 million between 2015 and 2022. A number of lobby groups like the Minerals Council of Australia, and the Australian Petroleum Production and Exploration Association are also regular and significant donors.[25]

Citing figures from the activist organisation Lock the Gate, the Climate Action Network further notes that companies receiving approvals from roughly 1998 to 2022 made a little under $55 million in political donations, maintaining that, 'they [fossil fuel companies] don't donate this money for nothing but to ensure fossil fuel subsidies are not phased out and to hinder changes to law or regulation that hinder fossil fuel extraction'.[26] This seems a reasonable conclusion to draw given the desire of fossil fuel companies to have their project applications approved. It speaks volumes, for instance, that Woodside and Chevron, leading Australian oil and gas companies, according to advocacy organisation Renew Economy, donated $55,600 and $52,500 respectably to the Labour party, 'during a period when the Albanese Government was making crucial decisions about the expansion of Australia's oil and gas industries – and moving to introduce price caps on gas supplies in an effort to lower prices for consumers'.[27] In a 2023 report by the energy analytics organisation Market Forces titled *Dirty and Dark – Political donations from the fossil fuel industry*, it was found that despite dire

---

24 Market Forces, *Dirty and dark: political donations from the fossil fuel industry* (1 February 2023, Market Forces), <https://www.marketforces.org.au/politicaldonations2023/>, accessed 6 April 2025.

25 Market Forces, *Dirty and dark*.

26 Climate Action Network Australia (CANA), *Dirty power: political donations from the fossil fuel industry* (2022, CANA), <https://www.cana.net.au/dirty_power_report>, accessed 6 April 2025.

27 Bernard Keane, 'Tanya Plibersek approves Santos gas project', *Crikey* (22 February 2023), <https://www.crikey.com.au/2023/02/22/tanya-plibersek-approves-santos-gas-project/>, accessed 6 April 2025; Michael Mazengarb, 'Millions in fossil fuel money rained on major political parties last year', *Renew Economy* (1 February 2024), <https://reneweconomy.com.au/millions-in-fossil-fuel-money-rained-on-major-political-parties-last-year/>, accessed 6 April 2025.

warnings about emission levels from various climate monitoring organisations, the current Albanese Labour government has given the green light to numerous new coal and gas projects. Six new coal mining projects were approved in the first few months of the newly installed administration, with 118 more coal, oil and gas projects in the investment pipeline.[28] Many of these projects have since been given the go-ahead even as the government seeks a 62–70 per cent reduction in Australia's emissions by 2035.[29] Such figures reveal not isolated policy failures but the systemic capture of the state by extractive interests.

Of particular concern here, is the role played by highly remunerated government lobbyists and think tanks which both actively support and promote the dirty energy sector.[30] Many of these lobbyists are former senior politicians or public servants who move seamlessly through the revolving door between government and industry, leveraging their insider knowledge and networks to influence policy decisions in favour of fossil fuel interests. Industry-aligned think tanks – such as the Institute of Public Affairs and the Centre for Independent Studies – further reinforce this influence by shaping public discourse, publishing reports, and appearing in the media to question or downplay the urgency of climate action. Together, these actors sustain an echo chamber of economic rationalism in which fossil fuel expansion is framed as essential for 'national prosperity' and 'energy security'. Their interventions normalise ongoing extraction and delay structural reform, effectively embedding corporate interests within the machinery of governance.

While it is difficult to prove a direct link between donations and particular project approvals, the fact that, for example, the Albanese government has given

---

28 The Climate Council here sets out the number of new approvals in 2025 alone: Climate Council, *New analysis: the Albanese government's fossil fuel approvals* (3 September 2025, Climate Council), <https://www.climatecouncil.org.au/resources/albanese-government s-fossil-fuel-approvals>, accessed 6 April 2025.

29 In September 2025, the Australian Federal Government introduced new emissions reductions targets of 62–70 per cent reductions on 2005 levels. Journalist Tony McIllroy notes that the previous ER target is not on track to be achieved. Tom McIlroy, 'Labor's 2035 emissions target may be middle of the road but it shows there's no going back on climate', *The Guardian* (19 September 2025), <https://www.theguardian.com/environment/2025/sep/19/labors-203 5-emissions-target-may-be-middle-of-the-road-but-it-shows-theres-no-going-back-on-climate>, accessed 6 April 2025.

30 Comms Declare, *Fossil fuel lobbyists flood Canberra as agencies and media exposed in 2025 F-list* (25 September 2025, Comms Declare), <https://commsdeclare.org/2025/09/25/fossi l-fuel-lobbyists-flood-canberra-as-agencies-and-media-exposed-in-2025-f-list/>, accessed 15 October 2025; See also: Mazengarb, 'Millions in fossil fuel money rained on major political parties last year'.

the go-ahead to companies that have made donations should be a matter of great public concern. Arguably, this is yet another brick in the hegemonic wall that enables fossil fuel companies to carry on with their polluting operations.

Financial complicity

Fossil fuel companies are further supported by the financial sector, particularly banks, which over recent decades have bought into the profit maximisation agenda. In July 2024, Market Forces reported that since the 2015 Paris Agreement drawn up by 194 countries plus the European Union to limit climate change to below 2 degrees, Australia's big four banks, ANZ, NAB, Commonwealth Bank of Australia and Westpac loaned fossil fuel companies more than AU$61 billion. In 2023, these banks poured $3.6 billion into fossil fuel projects, 'including $2.5 billion in funding for companies expanding the coal, oil and gas industries, despite all four banks committing to global climate goals'. In the same year, however, Australian banks for the first time did not invest directly in new or expanding fossil fuel projects, although they did use general purpose corporate lending and bonds as 'backdoor financing options'. The Big Four, Market Forces found, 'continued greenwashing by providing almost 70 per cent of their fossil fuel lending to companies out of line with their climate commitments'.[31]

The 'monumental façade' created by these banks, renders them directly culpable in worsening biospheric conditions.[32] They are joined by seven other Australian banks and eight subsidiaries, investing heavily in Australian fossil fuel companies, according to Market Forces.[33] Additionally, foreign banks like the Bank of China, HSBC and ING continue to invest heavily in the sector, amounting to $66.2 billion in 2023 alone.[34]

Another sector responsible for funding polluting companies is superannuation. In 2024, Market Forces reported that, 'Australia's top 30 super funds have more than $39 billion invested in companies globally with the biggest fossil fuel expansion plans'.[35] The majority of these funds, which have doubled over a two-year

---

31 Market Forces, *Banking climate failure: big four Australian banks pour $3.6 billion into fossil fuels in 2023* (16 July 2024, Market Forces), <https://www.marketforces.org.au/media-release-banking-climate-failure-2024/>, accessed 6 April 2025.

32 Market Forces, *Banking climate failure*.

33 Market Forces, *Banking climate failure*.

34 Market Forces, *Banking climate failure*.

35 Market Forces, *Super funds invest $39 billion in fossil fuel growth* (28 May 2024, Market Forces), <https://www.marketforces.org.au/super-funds-invest-39-billion-in-fossil-fuel-growth/>, accessed 6 April 2025.

period, flow into 190 companies who feature prominently in 'The Climate Wreckers Index'.[36] Worryingly, Market Forces notes a decline among the top 30 superfunds of investments going into clean energy projects. As Superannuation Funds Campaigner at Market Forces, Brett Morgan, remarked: 'Investments in the world's biggest climate wreckers are skyrocketing as Australia's biggest super funds are failing to rein in dangerous coal, oil and gas growth'.[37] Three companies, Woodside Energy, Santos and Whitehaven Coal, according to Market Forces, are, 'responsible for 59 per cent of projected emissions attributable to the fossil fuel expansion plans of companies'. The leading super-offenders investing in fossil fuel expansion are UniSuper, Commonwealth Super Corp, and MLC.[38]

Despite claims that they remain committed to the growth of clean energy, Australian banks and superannuation funds play a key role in propping up industries whose products are directly injurious to planetary health and wellbeing. Shareholder backlash and other activist opposition, has not prevented fifteen of Australia's leading superannuation funds from holding over $25 billion in fossil fuel assets, as of 25 August 2025.[39]

The media

Across the media landscape, coverage of climate-related issues has increased significantly over recent years – although not in ways that fully inform the public of actual risk levels. In a 2024 report titled *Climate Change Reporting in the Australian Media*, a multidisciplinary team of researchers at RMIT's Innovation Catalyst initiative found that the number of climate change articles published annually in the 1990s was around 500-plus, whereas by the 2010 this had ballooned to over 80,000 per year.[40] Despite this huge increase, the researchers found that the number of news items focused on climate trends and patterns presented by organisations like the IPCC had fallen dramatically, with attention largely given over to more 'newsworthy' live-streamed climate events. Moreover, proportionately few articles focused on regenerative projects, community-based climate initiatives or alternatives to fossil fuel. On another front, the report found that

36 Market Forces, *Super funds invest $39 billion*.
37 Market Forces, *Super funds invest $39 billion*.
38 Market Forces, *Super funds invest $39 billion*.
39 Market Forces, *Super funds invest $39 billion*.
40 RMIT Innovation Catalyst and ANU Global Institute for Women's Leadership, *Climate change reporting in the Australian media: data for good insights paper* (November 2024, RMIT), <https://giwl.anu.edu.au/files/2024-11/ssrn-4851963.pdf>, accessed 6 April 2025.

coverage of climate protests lessened over time, especially after the high point of 2019 – protests and the reason for them, were no longer deemed newsworthy. Overall, the RMIT team found that the 24/7 news cycle and focus on ratings tended to dictate how climate-related events were repeated. This is borne out by Deakin University researchers Gabi Mocatta and Erin Hawley who note that there was a spike in news coverage on climate issues around the time of the 2019 bushfires which then slowly ebbed away until the 2022 floods, which garnered considerable media attention.[41] In such instances, the media tended to dwell on the consequences of extreme weather events – the actual fires, floods, physical damage, accounts of victims etc – with only scant attention paid to climate trends – a practice long evident in media news coverage, as journalist Jessica Aldred observed back in 2013 following the 'angry summer' of that year.[42] The complex political and environmental causes of these events, and the longer-term trauma experienced by victims, was of less interest to media outlets than the immediate and 'spectacular' nature of the events themselves.

The Covid-19 pandemic in 2020 had a particularly marked effect on media coverage of climate issues, which fell by more than a third, according to a multi-university team of researchers at the Media and Climate Change Observatory.[43] It was noted that the total number of climate-related stories in early 2020 dropped sharply across major Australian newspapers – including *The Age*, *The Australian*, *Sydney Morning Herald*, *Courier Mail*, and *Daily Telegraph* – a decline more pronounced than that seen in the UK or US. Coverage recovered somewhat late in 2020 with more climate-related reports as COP26 approached. The patchy coverage of the year 2020, however, tended to conceal what the World Health Organisation said was as a 'double blow' for millions of people as a result of both Covid and extreme weather events.[44] The year featured record-breaking Atlantic hurricanes (30 in all), huge forest fires in Australia, California, Siberia,

---

41 Grahame Mocatta and Emily Hawley, 'Uncovering a climate catastrophe? media coverage of Australia's Black Summer bushfires and the revelatory extent of the climate blame frame', *M/C Journal* 23/4 (2020), <https://doi.org/10.5204/mcj.1666>.

42 Jessica Aldred, 'Australia links "angry summer" to climate change – at last', *The Guardian* (7 March 2013), <https://www.theguardian.com/environment/blog/2013/mar/07/australia-angry-summer-climate-change>, accessed 6 April 2025.

43 Ami Nacu-Schmidt, Olivia Pearman, Jennifer Katzung and Max Boykoff, '*This historic decline in emissions is happening for all the wrong reasons…*', *MECCO Monthly Summaries*, Issue 40 (May 2020), <https://scholar.colorado.edu/concern/articles/wm117p983>, accessed 6 April 2025.

44 World Health Organization (WHO), *Climate change and health* (2023, WHO), <https://www.who.int/news-room/fact-sheets/detail/climate-change-and-health>, accessed 6 April 2025.

and the Amazon, and foods and cyclones in South Asia, East Africa, and Southeast Asia. Prolonged heatwaves also occurred in the US, Europe, Siberia and the Arctic, and there were droughts in Southern Africa and parts of South America. While Covid-19 was an understandable distraction, the impacts of the climate disaster were still playing out across the world – even though these were largely ignored by much of the media.[45]

The general lack of detailed attention given to climate issues, inducing the climate science, has been further compromised by dis/misinformation campaigns, with NewsCorp playing a key role in this regard. Following the resignation in 2023 of Rupert Murdoch, the climate-sceptic Chairman of News Corp and Fox News, Dr Joëlle Gergis remarked that: 'It's hard to think of another person who has single-handedly done more to muddy the public's understanding of climate change', adding: 'we have wasted decades debating the fundamental science in the media, when we really should have been focused on urgently implementing climate policies that will genuinely reduce emissions'.[46] The dis/misinformation campaign prosecuted by NewsCorp through *The Australian* and other outlets like the *Daily Telegraph*, *Herald Sun* and TV station, *Sky News*, according to journalists Readfearn and Adam Morton, had a detrimental effect on public responses to the climate science, sowing doubt and scepticism over the credibility of the climate data and climate scientists.[47] Through selective storytelling, the media both reveal and obscure the crisis, creating a sense of urgency without consequence – a hallmark of hypernormalisation.

As Kate Walton noted in a 2020 article for *Al Jazeera*, the Murdoch press, even in the wake of the catastrophic 2019 Australian bushfires, was playing down or denying the climate science, going as far as describing climate activists as 'climate catastrophists' and part of a 'socialist plot'.[48] Despite what appeared to be a conversion to the climate science in 2021 (probably in response to mounting public concern over numerous climate disasters), it appears, according to

[45] World Meteorological Organization, *State of the global climate 2020* (Geneva: World Meteorological Organization, 2021), <https://wmo.int/media/news/climate-change-indicators-and-impacts-worsened-2020>, accessed 25 November 2025.

[46] Graham Readfearn and Adam Morton, '"climate villain": scientists say Rupert Murdoch wielded his media empire to sow confusion and doubt', *The Guardian* (23 September 2023), <https://www.theguardian.com/media/2023/sep/23/rupert-murdoch-climate-change-denial>, accessed 25 April 2025.

[47] Readfearn and Morton, 'climate villain'.

[48] Kate Walton, 'how "murdochracy" controls the climate debate in Australia', *Al Jazeera* (24 January 2020), <https://www.aljazeera.com/news/2020/1/24/how-murdochracy-controls-the-climate-debate-in-australia>, accessed 25 April 2025.

University of Adelaide academic Victoria Fielding, that News Corp's penchant for climate scepticism remains a feature of its ongoing coverage.[49] Given that NewsCorp controls around 65 per cent of the Australian newspaper market and commands a significant share of both cable and free-to-air television audiences, it is unsurprising that climate policy has remained conflicted and sluggish – at least until the new government came to power in 2022. Up to that point and beyond, activist groups and NGOs have led efforts to hold fossil fuel companies and their enablers to account.

Energy companies like ExxonMobil and many of the scientists they employed and originations they funded, have done what they can to distract attention from the climate science. As academics Naomi Oreskes and Erik Conway point out in *Merchants of Doubt: How a Handful of Scientists Obscured the Truth on Issues from Tobacco Smoke to Climate Change*, fossil fuel companies have, over many years, actively funded and disseminated misinformation through various public channels, depicting the science as questionable, and a threat to the economy. This despite the fact that, as early as the 1950s, they supported research linking carbon dioxide emissions to global warming – research that has consistently confirmed the risks posed by their own products.[50]

Writing in *The Conversation*, US historian Benjamin Franta reviewed internal documents from major fossil fuel companies such as Exxon (US), Shell (Netherlands), and Total (France), revealing that these corporations have long been aware of the dangers posed by greenhouse gas emissions.[51] Despite possessing such knowledge, they have actively sought to suppress their findings and mislead the public and official inquiries. Suffice to say, they also continued to pursue their harmful extractivist operations, which over the years morphed into aggressive 'greenwashing' where clever marketing techniques masked dangerous industrial practices. Nowadays, as Franta observes, '... most oil companies [including Australian companies like Chevron, Woodside and Santos] shy away from denying climate science outright, but they continue to fight fossil fuel controls

49 Victoria Fielding, 'Is News Corp following through on its climate change backflip? my analysis of its flood coverage suggests not', *The Conversation* (25 March 2022), <https://theconversation.com/is-news-corp-following-through-on-its-climate-change-backflip-my-analysis-of-its-flood-coverage-suggests-not-179468>, accessed 25 April 2025.

50 Naomi Oreskes and Erik M. Conway, *Merchants of doubt: how a handful of scientists obscured the truth on issues from tobacco smoke to climate change* (New York: Bloomsbury Press, 2010).

51 Benjamin Franta, 'Big oil knew: documents reveal fossil fuel industry's decades of climate knowledge', *The Conversation* (25 October 2021), <https://theconversation.com/big-oil-knew-documents-reveal-fossil-fuel-industrys-decades-of-climate-knowledge-170227>, accessed 25 October 2025.

and promote themselves as clean energy leaders even though they still put the vast majority of their investments into fossil fuels'.[52] This sort of deception has been exposed by some sections of the Australia media. Other mainstream outlets however, are heavily compromised by the advertising revenue they continue to receive from fossil fuel companies. In 2024, it was estimated that, 'Australia's top five fossil fuel advertisers spend more than $200 million annually on advertising across all platforms including radio, billboards, and digital'.[53]

This is not innocuous. Gergis recalls that according to the most recent data from the Global Carbon Project, around 90 per cent of global carbon dioxide emissions in 2024 came from the burning of fossil fuels, with the remaining 10 per cent linked to land-use changes such as deforestation and wildfires.[54] The rise in fossil fuel emissions that year was largely driven by growing gas and oil consumption, which together accounted for just over half of global emissions. Meanwhile, coal use reached a record peak, responsible for approximately 41 per cent of total emissions.[55] Despite the outcomes of COP28 urging nations to accelerate the transition away from fossil fuels, current production plans indicate a stark contradiction: by 2030, governments and corporations still intend to extract more than twice the volume of fossil fuels compatible with the 1.5°C target.[56] As Gergis explains, this ongoing expansion is often justified by framing gas as a 'transition fuel' with supposedly lower emission intensity, masking its continued role in deepening the climate crisis.[57]

AI also plays an increasingly important role in shaping how people perceive the world, reinforcing the dynamics that sustain hypernormalisation. Algorithmic systems curate what appears in newsfeeds and search results, filtering the overwhelming complexity of contemporary life into personalised streams of content designed for attention rather than understanding. This selective visibility makes it harder to grasp the structural dimensions of the climate crisis. By amplifying distraction, narrowing focus, and smoothing over contradiction, AI-driven media environments help maintain the appearance of stability even as ecological breakdown accelerates. The result is not overt censorship but a subtle, automated narrowing of the imaginable, a constant gentle pressure that keeps

---

52 Franta, 'Big oil knew',

53 Bill Noble, 'Should Australia ban fossil fuel advertising?', *The Australia Institute* (7 July 2024), <https://australiainstitute.org.au/post/should-australia-ban-fossil-fuel-advertising/>, accessed 25 October 2025.

54 Gergis, 'Net zero distracts from scientific imperative to eliminate fossil fuels'.

55 Gergis, 'Net zero distracts from scientific imperative to eliminate fossil fuels'.

56 Gergis, 'Net zero distracts from scientific imperative to eliminate fossil fuels'.

57 Gergis, 'Net zero distracts from scientific imperative to eliminate fossil fuels'.

public consciousness oriented toward the familiar and away from the systemic forces that require disruption.

Universities

Educational institutions play a pivotal role in sustaining fossil fuel interests. In Australia, as elsewhere, this link is often opaque and subsumed under the carefully manufactured image of universities as ardent proponents of renewable energy. As we note below, the reality is much more complex.

The ambivalence displayed by universities is demonstrated in a number of different ways. For example, it is evidenced in the lack of interest shown by academics about the role played by fossil fuel companies in their own and other institutions. In May 2024, journalist Dhamma Noor reported on a major international study in the journal *WIREs Climate Change* which, among other things, noted a distinct lack of interest among Australian academics in the links between higher education research funding and the fossil fuel industry.[58] Of the roughly 14,000 peer-reviewed articles dealing with the influence of gas and oil companies in various sectors between 2003 and 2023, only a handful dealt with the higher education sector. This self-imposed censorship (or blind spot?) helped conceal the fact that energy companies were investing heavily in university-based climate and energy research, and that industry representatives held positions on numerous advisory and governance boards. Additionally, the industry was, 'endowing academic posts, sponsoring scholarships, advising curricula or otherwise influencing universities'.[59] This, says Noor, is reminiscent of the influence exercised by pharmaceutical and tobacco industries in tertiary institutions in the US and elsewhere.

In 2023, journalist James Gallagher documented the incursions of fossil fuel companies across a range of Australian higher education activities. These included: 'Open, blatant partnerships' where mining companies recruited university-educated graduates and promoted research focused on how 'to exploit Earth's natural resources most efficiently, and… use some of their ample profits to pay for it'.[60] The University of Sydney, for example, partnered with Rio Tinto

---

58 Dhriti Noor, 'University funding from fossil fuels slowing switch to green energy – report', *The Guardian* (5 September 2024), https://www.theguardian.com/business/article/2024/sep/05/universities-fossil-fuel-funding-green-energy>, accessed 6 April 2025.

59 Noor, 'University funding from fossil fuels slowing switch to green energy'.

60 James Gallagher, 'Partners in pollution: Australian universities and the fossil fuel industry', *Redflag* (9 May 2023), <https://redflag.org.au/article/partners-pollution-australian-universities-and-fossil-fuel-industry>, accessed 6 April 2025.

in overseeing the Rio Tinto Centre for Mine Automation, while the University of Western Australia partnered with leading coal and gas company, Woodside. Additionally, the government, through the Australian Research Council, has handed out 'significant amounts of public money to projects that benefit the fossil fuel industry'.[61] Mining companies also offer direct industry funding to promote university research. An example of this is the Australian Coal Industry's Research Program (ACIRP), a 'slush fund' entirely owned and funded by Australian black coal producers, which has generated $91 million funding for 277 mostly partnered research projects across 15 Australian universities. The ACIRP also receives applications for research funding which is overseen by coal executives. The universities of Queensland, Newcastle and New South Wales are leading recipients. Mining companies also offer endowments, scholarships, internships, research grants and hold careers expos, all of which binds them ever closer to universities.[62] These links are strengthened in other ways too, mostly through the involvement of fossil fuel executives in governance positions: 'Almost every Australian university', Gallagher finds, 'has a board member with a background in fossil fuels'.[63]

Australia is not alone in such entanglements. As the authors of the *Wires* report conclude: 'In an era of widespread concern about corporatization and declining public funding of higher education, and a worsening climate crisis met by limited government action, fossil fuel companies have embedded themselves in universities across the US UK, Canada, Australia, and beyond ... Our review suggests that universities are an established yet under-researched vehicle of climate obstruction by the fossil fuel industry, complementary to its documented history of climate denial, delay, and lobbying'.[64]

Who bears responsibility for the climate disaster?

In June 2020, freelance science journalist Jocelyn Timperley put the question: 'Who is really to blame for climate change?'[65] It's an inquiry that leads us down some interesting paths. To be sure, fossil fuel companies, through their duplicity and continued commitment to investing in extractivist operations, are easy targets. And rightly so. Clearly, these companies have a vested interest in averting

61 Gallagher, 'Partners in pollution'.
62 Gallagher, 'Partners in pollution'.
63 Gallagher, 'Partners in pollution'.
64 Quoted in Noor, 'University funding from fossil fuels slowing switch to green energy'.
65 Jocelyn Timperley, 'Who is really to blame for climate change?', *BBC Future* (19 June 2020), <https://www.bbc.com/future/article/20200618-climate-change-who-is-to-blame-and-why-does-it-matter>, accessed 6 April 2025.

the public's gaze from the source of the problem. One of the industry's repeated observations is that without demand for energy-intensive consumer products there would be no climate problem. In other words, it is the consumer who is to blame. This argument has been debunked by Jeff Sparrow in *Crimes against Nature: Capitalism and Global Heating*.[66] Sparrow argues that the fossil fuel industry's narrative of 'shared responsibility' is one of the most effective ideological manoeuvres of late capitalism. By framing the predicament as the cumulative result of individual choices – our driving habits, dietary preferences, or household energy use – attention is deflected from the structural forces that produce and profit from ecological destruction. He reminds us that consumers did not design the energy system, set carbon-intensive supply chains in motion, or determine the economic logic that privileges endless growth over planetary survival. Rather, it is corporations, financial institutions, and complicit governments that have locked societies into fossil dependency while marketing consumption as freedom. In this light, personal responsibility becomes a convenient smokescreen – one that obscures corporate culpability and maintains the illusion that systemic transformation can be achieved through lifestyle adjustments alone.[67] The industry's disingenuous attempt to present itself as morally benign is part of a deliberate strategy to ensure the status quo. In this sense, consumer guilt operates as a form of hegemonic consent, ensuring that citizens internalise responsibility for a system over which they have little real control.

While it is arguable that citizens as consumers are all co-responsible for many of the vagaries of extractivist capitalism – we do, after all, purchase fossil fuel generated products and services – this view obscures the full picture. As citizens and consumers, we are enmeshed in a system in which it is easier to go along with a way of life that meets our daily needs than to forge a more morally resonant path in line with our values. In short, our ability to make the 'right', informed choices, is significantly affected by how our society is organised and experienced. Mass consumption, for example, has evolved through a process of enculturation involving many decades of marketing and promotion in which the acquisition of goods and services is linked to class, status and identity.

Capitalism does not merely produce goods and services; it produces subjects – individuals whose desires, aspirations, and sense of self are moulded through the logics of consumption. From early childhood, citizens are socialised into an economy of wanting, where identity and belonging are tied to what one owns,

---

66 Jeff Sparrow, *Crimes against nature: capitalism and global heating* (Melbourne: Scribe Publications, 2021).

67 Sparrow, *Crimes against nature*.

wears, or displays. Advertising, media, and digital technologies continuously cultivate this consumer self, presenting consumption as both a moral and emotional act – an expression of freedom, success, even care for the planet through the language of 'ethical' or 'green' products. In this way, capitalism transforms structural dependency on fossil-fuelled systems into a personal lifestyle choice, masking the deeper coercion at play. This process of psychological and cultural conditioning ensures that the dominant economic order reproduces itself not only through institutions and policies but through the everyday thoughts and behaviours of its participants. The 'consumer', far from being an autonomous decision-maker, is a product of this system – trapped within an infrastructure and ideology that constrains meaningful alternatives. It is too easy – and convenient – to point the finger at 'consumers' without fully comprehending how capitalism itself shapes human consciousness in the context of a competitive market society. The challenge, then, is not to moralise individual consumption but to confront the social and economic machinery that makes high-carbon living appear desirable and inevitable.

The behavioural dimension of this system is further intensified by AI-driven surveillance capitalism. Digital platforms now collect and analyse vast amounts of personal data to anticipate and influence individual preferences, emotional states and purchasing decisions. These algorithmic systems do more than respond to consumer desire – they actively cultivate it. By steering attention, shaping habits, and personalising advertising with extraordinary precision, they deepen the link between identity and consumption. Choices that appear personal or expressive are frequently the outcome of engineered digital environments designed to maximise engagement and expenditure. This produces a consumer subject who feels responsible for the crisis yet has limited meaningful agency within an economy that continually nudges behaviour toward high-carbon lifestyles. AI thus reinforces the illusion of individual responsibility while obfuscating the structural forces that profit from these patterns.

As climate scholar and author Genevieve Guenther observes: 'The *we* responsible for climate change is a fictional construct, one that's distorting and dangerous', adding that: 'By hiding who's really responsible for our current, terrifying predicament, political cover [is provided] for the people who are happy to let hundreds of millions of other people die for their own profit and pleasure'.[68] Further, as George Monbiot observes: 'While we are all participants in the first

68 George Monbiot, 'The big polluters' masterstroke was to blame the climate crisis on you and me', *The Guardian* (9 October 2019), <https://www.theguardian.com/commentisfree/2019/oct/09/polluters-climate-crisis-fossil-fuel>, accessed 7 April 2025.

great extermination, our responsibility is not evenly shared. The impacts of most of the world's people are minimal. Even middle-class people in the rich world, whose effects are significant, are guided by a system of thought and action that is shaped in large part by corporations'. The attempt to attribute blame for our collective predicament to individual consumers, is, says Monbiot, grossly misleading. 'This individuation of responsibility, intrinsic to consumerism, blinds us to the real drivers of destruction'.[69]

In its December 2023 report titled *Who is responsible for climate change?* Oxfam identifies the global drivers as: 'Big Oil, rich industrialised countries, and carbon billionaires', with the bulk of carbon emissions 'traced to just 100 fossil fuel producers'.[70] Twenty-three of the leading industrialised countries, says Oxfam, accounted for half of all historical emissions, with industrialised nations in Europe, North America, Australia, and Japan responsible for over three-quarters of all global emissions between 1751 and 2006.[71]

Between 1990–2015, Oxfam continues, 'the carbon emissions of the super-rich globally were more than double the emissions of the poorest half of humanity. At the same time, the poorest 50 per cent – around 3.1 billion people – were responsible for just 7 per cent of emissions'. Wealthy polluters, and the companies they own or invest in, Oxfam maintains, are, 'responsible for climate change – and it's time to hold them accountable on the world stage'.[72]

For climate watchers, it is galling to witness the enormous profits made by fossil fuel companies at the expense of some of the world's poorest nations. In 2023, the Climate Council reported that the three biggest Australian-based fossil fuel companies made gargantuan profits while emitting huge amounts of toxic emissions.[73] Leading the pack was Chevron Australia, described by the Council as 'a notorious tax avoider and polluter', producing more than 52 million tonnes of carbon pollution or around 9 million tonnes per year.[74] Between 2020 and 2021, Chevron made $12.5 billion in total revenue from selling Australia's resources, yet paid only $30 dollars in company tax. In 2023, the company made around

---

69 Monbiot, 'The big polluters' masterstroke'.

70 Oxfam, *Who is responsible for climate change?* (3 December 2023), <https://www.oxfamamerica.org/explore/stories/who-is-responsible-for-climate-change/>, accessed 7 April 2025.

71 Mohamed Adow, 'The climate debt: what the West owes the rest', *Foreign Affairs* (May/June 2020; published online 13 April 2020), <https://www.foreignaffairs.com/articles/world/2020-04-13/climate-debt>, accessed 7 April 2025.

72 Oxfam, *Who is responsible for climate change?*. This practice of self-enrichment at the cost of poorer nations has been referred to as 'carbon colonialism'. Parsons, L, 2023, Carbon *Colonialism How rich countries export climate breakdown*, Manchester, MUP.

73 Oxfam, *Who is responsible for climate change?*.

74 Oxfam, *Who is responsible for climate change?*.

$6 billion in profits. Woodside and Anglo American, second and third respectively in the revenue rankings, also made billions in profits from their operations, while also paying little or no company tax. Other leading Australian companies making significant profits include Santos, BHP, Glencore Coal, INPEX, Shell Australia and ConocoPhillips.[75]

The so-called 'dirty dozen' extractivist companies, according to the Climate Council, have emitted 287 million tonnes of CO2 equivalent emissions since 2016, around 10 per cent of Australia's total national carbon emissions.[76] They have done so, as noted, with the assistance of generous government subsidies, loans from banks, and investments by super funds.

The governments' continued support for the fossil fuel industry casts a long shadow over claims that Australia is an exemplary front runner in the fight against climate chaos.

Conclusion

We began this chapter with Cat's story, which ended with the question: '*Why are we pretending it's situation normal?*' The blunt answer is that this illusion serves the strategic and commercial interests of powerful actors and is sustained by a system that depends on collective denial to perpetuate itself. Put bluntly: profit margins rule the day. Despite harrowing evidence on the disastrous effects of a warming planet, it appears that fossil fuel companies and the billionaire class, in cahoots with governments, financial institutions, universities and the corporate media, remain hell-bent on making a bad situation worse – a scenario described by professor emeritus of philosophy at University of Pennsylvania, Wendy Lynne Lee, as 'eco-nihilism'.[77]

'Business as usual' in this sense is not a neutral condition but something akin to an induced mass dissonance – a psychological state requiring constant reinforcement through all the hypernormalised techniques available to the powerful. The dominant order, aptly described as 'cannibal capitalism',[78] thrives on

---

75 Oxfam, *Who is responsible for climate change?*.

76 Climate Council, *Introducing the dirty dozen: Australia's filthiest fossil fuel polluters* (Climate Council, February 2023), <https://www.climatecouncil.org.au/wp-content/uploads/2023/02/Climate-Council-Report-Introducing-the-Dirty-Dozen-Feb-2023-1.pdf>, accessed 7 April 2025.

77 Wendy L. Lee, *Eco-nihilism: the philosophical geopolitics of the climate change apocalypse* (Lanham, Pennsylvania: Lexington Books, 2017).

78 Nancy Fraser, *Cannibal capitalism: how our system is devouring democracy, care, and the planet – and what we can do about it* (London: Verso Books, 2022). Self-cannibalism, says ecologist Bill McKibben, is capitalism's suicidal intent. Bill McKibben, 'Capitalism is acting truly

its capacity to turn disaster into opportunity, grief into growth, and destruction into dollars. The very disasters it causes become pretexts for further extraction and technological 'solutions' that deepen dependence on the same system.

As we have seen, this co-dependent machinery is not upheld by corporations alone. It is reinforced by governments that subsidise and legitimise fossil fuel expansion, financial institutions that bankroll destruction while greenwashing their investments, universities that lend credibility through partnership and silence, and media organisations that neutralise threat to preserve a sense of order. Each of these sectors participates in the choreography of hypernormalisation – sustaining the illusion that our predicament is manageable, reversible, and compatible with ongoing economic growth.

At the ideological level, the system's resilience lies in its capacity to reshape consciousness itself. By individualising responsibility and rebranding consumption as moral choice, it obscures the structural drivers of collapse and fragments collective agency. The consumer is recast as both culprit and saviour, while those who profit most from chaos remain shielded by narratives of necessity, progress and innovation.

Yet each flood, fire, and heatwave exposes the limits of this delusion, eroding the legitimacy of institutions that insist on business as usual. As the myth of stability unravels, new possibilities emerge – for truth-telling, solidarity, and reimagining what it means to live well on a damaged planet. To name the system's pathology is itself an act of resistance, for it shatters the spell of inevitability that sustains its power.

If hypernormalisation is the condition of our age, then breaking it requires a counter-practice: civic wakefulness. It means confronting the grief, anger, and dissonance that the system compels us to suppress, and transforming these into political and social energy. The challenge ahead is not only technological or economic but profoundly cultural and ethical – how to reclaim our capacity to feel, to imagine, and to act together outside the constraints of a dying system.

The dissolution of the old order provides spaces in which new forms of solidarity and meaning can take place. The rest of this book turns to these openings, asking how communities, movements, and institutions might move beyond illusion toward the hard work of collective transformation.

---

suicidally on climate', *Common Dreams* (18 April 2024), <https://www.commondreams.org/opinion/capitalism-suicidal-climate>, accessed 7 April 2025.

PART II

Reimagining Life as We Fall

Part I of this book invited us to face climate disruption and collapse with honesty and courage – to recognise the Great Unravelling not as a distant threat, but as the condition of our time. Having glimpsed the storm, we now turn to the vexed question of how shall we live in this time of endings?

Given the terrifying scenarios outlined in earlier chapters, it may seem a little trite to now explore our inner worlds and the importance of community resilience. After all, how do we go from chapters on collapse, climate disaster and hypernormalisation to talk about inner peace in Chapter 4? It feels like quite a leap. First, we argue that how we attend to our emotional and spiritual responses to what is unfolding matters – to our survivability and potential for meaningful existence. Second, we may have little choice but to ask what courage, care, and renewal look like when our world is falling apart. We will all have to contend with some very big ontological questions.

Part II therefore asks: what forms of meaning, connection, and practical action are possible amid breakdown? How can individuals and communities find steadiness and solidarity in such circumstances? How might we reimagine the systems that sustain and nurture life, rather than destroy it?

The liminal space we now inhabit – between resignation and renewal – requires us to respond with humility and care if we are to live well, or as well as we can, even in the midst of chaos and confusion. The next three chapters focus on agency – not the tokenistic sort celebrated in self-help books or policy discourse, but a grounded, collective capacity to act meaningfully from where we are. Where Chapter 3 examined how systems of power sustain the performance of normality even as ecological collapse accelerates, Part II turns to what interrupts this mirage. Agency, in this context, is not the illusion of control within a failing system, but the capacity to refuse its terms – to reduce harm, withdraw consent, and build alternatives that do not depend on denial. We ponder what each of us can do, here and now, without waiting for governments or institutions to lead.

This is not about offloading responsibility, but recognising that communities play a leading role in disaster response, and that strong social connections are the backbone of a resilient civic society.

It is in this context that we can learn to use our hands, our voices, our relationships, and our imaginations to shape what comes next. Sharing skills, caring for neighbours, growing food, joining local networks, Caring for Country or speaking up for what matters – these are all expressions of renewed solidarity: a form of power growing in cooperation, care, and the refusal to give up on the future. In suggesting pathways to personal peace and civic renewal, we offer neither utopian 'solutions', rescue, nor a return to normality. Instead, we propose ways of being that allow life to continue with dignity and purpose – through the cultivation of radical acceptance and deep peace, the rebuilding of civic life and neighbourly ties, and ultimately, the fulfilment of collective transformation.

Guided by the ethos of mutual aid – non-utopian networks of care and cooperation – we can learn to mitigate harm, adapt to change, grow resilience, and regenerate the conditions for a life worth living. In doing so, we can shift from fear to participation, from despair to meaning – even though the process may be unpredictable and messy. We can learn, too, that grief can coexist with gratitude, and that collective action and change can arise from inner stillness rather than panic.

Throughout this book, the term Country is used in its First Nations sense, where it refers to more than land alone. Country encompasses people, place, waters, skies, ancestors, stories, and law, understood as an interconnected and living whole. Within this worldview, humans are not positioned as external observers or owners of land, but as participants within a web of relationships that carries obligations as well as belonging. This understanding contrasts with dominant industrial frameworks in which land is treated primarily as resource or property. As Robin Wall Kimmerer, a celebrated Potawatomi botanist, scientist and decorated professor, observes, Indigenous knowledge traditions describe human–Earth relationships as reciprocal rather than extractive, shaped by ongoing responsibilities to that which sustains life. Seen through this lens, environmental degradation can be understood not only as ecological damage, but as a disruption of complex inter-relationships. The references to care for place, land, community, and regeneration that appear throughout this book draw on this relational understanding of Country. They situate responses to climate disruption within particular places and relationships, rather than in the nebulous and abstract, and help explain why in this book practices of care, mutual aid, and custodianship recur are central themes.

Our invitation is to move beyond survival toward renewal – to rediscover the art of living well, together, on Country, in a radically reconfigured world. None of this is easy – or impossible.

Living with mortality: From despair to deep peace

It might be strange to start a story with an ending. But all endings are also beginnings. We just don't know it at the time

—MITCH ALBOM

My barn having burned down, I can now see the moon

—MIZUTA MASAHIDE

The basic paradox. Everything is a mess, yet all is well

—EZRA BAYDA

Introduction

To live in this century is to live in the shadow of endings. Species vanish, forests burn, rivers run dry, and weather patterns once thought stable, now lurch from one extreme to another. Mortality – once imagined as an individual horizon – now looms over the entire planet, encompassing ecosystems, cultures, and futures – a chilling legacy of the Anthropocene. Yet, as noted in Chapter 1, even as collapse accelerates, public engagement with the issue of 'climate change' can often segway into denial, distraction, or compulsory optimism. We are urged to stay hopeful, to believe in technological fixes, to carry on as if the Great Unravelling were a figment of the eschatological imagination.

This sort of erasure reflects the resistance many of us have to our own deaths, and to the idea of death itself. As Canadian grief literacy advocate Stephen Jenkinson writes in *Die Wise: A Manifesto for Sanity and Soul*, 'That is the great labor of the thing, to let your life be changed utterly by the news of our death when all and everything around you is counselling steady as she goes, business as usual, as normal a life as possible under the circumstances'.[1]

This chapter inquires into what it means to face mortality directly – not to despair, but to live truthfully amid the process of dying. If the end of the world as we know it gives rise to grief, despair, and anger, these are understandable, perhaps inevitable emotional reactions which at times can feel overwhelming. These are not pathologies to be urgently eliminated but signals of our deep entanglement with what is passing away. When acknowledged, however, as we argue throughout this chapter, such emotions can open space for radical acceptance: a willingness to remain present to reality, however unbearable that might feel, and to act from that position. This is, we contend, the foundation of renewal.

None of this means abandoning the struggle for justice and survival. To the contrary: acceptance grounds us in the courage needed to embark on our challenging journeys – journeys shared in solidarity with others. Through acceptance may come deep peace, which, in contrast to resignation, is a steadying calm that allows us to act without being consumed by fear or rage – even though such feelings will undoubtedly arise. From a place of deep peace flows a renewed sense of presence and interconnection which acknowledges that life is sustained through webs of relation, not atomised existence. Here, we trace the movement from anger to acceptance, and peace to inter-being as an ontological foundation for life in a transformed world.

Stillness in action

Bobbi Allan has spent decades sitting with people at the point where inner lives and collapsing ecologies meet. A mindfulness teacher, community educator and longtime activist – she could also be described simply as an elder – her work has grown out of the unsettling realisation that the world she grew up in is changing in ways that cannot simply be '*fixed*'. Over the years, her activism shifted from campaigning to a steadier reckoning with long-term environmental decline. As she reflects, there was a season when '*we thought we might save the world in our lifetime*'; now the work feels more like learning how to stay present and useful as things unravel.

The phrase that eventually named this orientation came to Bobbi almost as a gift. '*Stillness in action*' described the quiet, grounded quality she cultivates in herself and supports in others. It does not mean withdrawal from the world. For

[1] Stephen Jenkinson, *Die wise: a manifesto for sanity and soul* (Berkeley, California: North Atlantic Books, 2015), 115.

Bobbi, stillness is what makes action possible. It is the inner steadiness that allows Bobbi and others to face hard truths, feel the ache of what is being lost, and still respond with care. Rather than rushing toward solutions, she invites people to notice what is happening in their own bodies and in the world around them.

A practice Bobbi uses in workshops and retreats is the 'truth mandala', a simple ritual that gives shape to core emotions and helps people recognise their life-giving side. Four objects are used to represent these emotions. The stick represents anger, but also the passion for justice. Dry leaves symbolise sadness, and yet, '*we wouldn't be sad if we didn't love, so it's the love that comes forth, really*'. The stone holds fear, but also underlines courage – '*to speak our fears, because, it's not easy to speak our fears. But we need to speak the unspeakable, so when we're holding the stone about our fears, that's our courage to do that*'. And she also uses an empty bowl, '*which is when you just feel like you're kind of gutted, or you don't know where to go next. But when we hold it, we also know that it's emptied us out for the new to emerge*'. Naming these emotions in community reframes them as anger-for-justice, fear-as-courage, sorrow-as-love, and emptiness-as-space for the new.

Bobbi's climate feelings are not abstract. She speaks candidly of periods of deep sadness, especially when she thinks of her grandchildren and of young people and the Earth. At times the scale of destruction feels overwhelming. At other times she feels a fierce determination simply to '*keep going*', to contribute whatever she can to the fabric of community resilience and inner steadiness. She recognises anger too, and how easily it can slide into fantasies of revenge or blame. When she notices this in herself, she lets it move through without becoming carried away by it. Instead, she talks about cultivating a kind of '*strategic patience*': utterly alert, but not wasting energy in frantic motion.

Underpinning this work is a long apprenticeship in Buddhist practice and the kind of spiritual inquiry that refuses to look away from suffering. Joanna Macy, scholar of Buddhism, systems thinking and deep ecology, has been a key companion in this journey. Through direct connection and friendship, Macy's *Work That Reconnects* helped Bobbi bring together two strands of her life that had previously felt separate: the contemplative path and the activist path. The recognition that our '*pain for the world*' is inseparable from our love for it, gave her a language for what she was already witnessing in groups. Macy's counsel continues to guide her, affirming despair as a doorway into connection and courage.

Connection with Earth is not, for Bobbi, a metaphor. It is a lived, bodily experience that has steadied her through times of despair. She speaks of sitting quietly outdoors, letting her senses open, and realising that the more closely she pays attention, the more relational the world becomes: birds calling to each other, wind moving through trees, the feel of soil under her feet. This attention is closely tied to gratitude. Having grown up with a strong sense of the labour and vulnerability behind food and land – shaped in part by her father's relationship to farming – she tends to look at any built environment and think about the ground beneath it, and the histories it holds. Such habits of noticing make it harder to fold into abstraction or to see the Earth only as backdrop to human dramas.

She remains acutely aware that not everyone lives near bushland, rivers or open fields. Many of the people she teaches are in dense urban environments, where concrete and noise dominate. Her advice to them is disarmingly simple: start where you are. There is, she insists, always some way to be in relationship with the more-than-human world, '*even if it's a pot plant on your twenty-sixth floor balcony*'. Even tiny encounters – a street tree, a sliver of sky – can be enough. '*Connection doesn't require wilderness. It requires attention*', she says. She often cites writers like Claire Dunn, who speak of 'rewilding the urban soul', as reminders that wildness is not only a property of remote landscapes but a quality of attention we can cultivate in everyday life.

What guides Bobbi's reflections is a quiet trust in life itself, even amid collapse. This is not naïve optimism. It sits alongside a sharp awareness of systemic violence, political failure and ecological damage. But she returns again and again to the sense that life is larger than any particular civilisation, and that humans have always had to find ways to live through upheaval. In her own words, there is a need for '*trusting life enough to stay present to it*', to keep looking for '*green shoots*' of renewal in the cracks of a fraying world. For Bobbi, organisations like Plan C, local mutual aid networks, and small circles of practice and reflection are examples of such green shoots – modest, imperfect, but real.

In bringing Bobbi's story into this chapter on mortality, despair and deep peace, we see how one person has woven together spiritual practice, psychological insight and community organising into a way of life that faces endings with dignity and purpose. Her commitment to stillness in action suggests that the work of facing ecological collapse is not only about abstract analysis and policy making, but also the slow, patient cultivation of presence – breath by breath, moment by moment, relationship by relationship, in whatever places we happen to live.

The mind-body-spirit continuum

Grief and loss are an integral part of the evolving climate breakdown story. They are visceral, troubling emotional reactions to chaos and breakdown, influencing how people make sense of, embody and respond to the Great Unravelling.

Numerous empirical studies have documented the pervasive nature of climate-related grief and anxiety, especially among young people. Such responses reflect deep anguish over the upending of a way of life, as well as burning anger at the institutional betrayal, greed, and duplicity. Seen in this context, ecological grief is not a pathology but a rational, shared response to hubris, naivety and systemic failure.[2]

Such is the growing recognition of the emotional and spiritual toll of climate chaos that the World Health Organisation now urges mental health considerations to be integrated into climate policy. Naming grief, accepting it, normalising it – even honouring it – allows us to acknowledge its presence as a deeply human response. Shared grief also helps steer us away from merely 'coping', and locates the possibility of repair and renewal in a place of recognition and acceptance. We are reminded here of the words of Indian philosopher and spiritual guide, Krishnamurti, who long ago opined that: 'It is no measure of health to be well adjusted to a profoundly sick society'.[3] In the same way, we might assert that 'you are not insane, the system is!'

As with the process of the Great Unravelling, our embodied responses will not take a predictable course. That said, our felt sense of connection and belonging, the compassionate, caring support we offer and receive, and the stories we tell ourselves and each other, will determine the extent of our trauma and moral injury. One major obstacle which we may wish to overcome, especially in Western societies, is the pervasive nature of individualism, social disconnection and loneliness which makes people much more reliant on external mental health services than strangers and loved ones. The sad fact is that many of us have forgotten or simply no longer know how to 'reach out', 'be with' and otherwise empathise with others in pain – a problem enhanced by our immersion in the world of digital technologies.[4]

This contrasts with ancestral forms of relationality in which the 'whole person' is defined not through individual identity but a web of relations – with family, community, ancestors, land, and more-than-human beings. Balance and harmony with the spiritual, emotional and communal realms is expressed through dreams, ceremony, ancestral knowledge, land-based practices, as well as seasonal and ecological cycles. All actions, choices and decisions are shaped by a deep connection and commitment to ancestors and descendants. This stands in stark contrast to the atomised, materialist conceptions prevalent in modern societies.

---

2 Ashlee Cunsolo and N. R. Ellis, 'Ecological grief as a mental health response to climate change-related loss', *Nature Climate Change* 8/4 (2018), 275–81.

3 Krishnamurti Foundation Trust (KFT), 'Regarding the quote "It is no measure of health…"', [no date], available at <https://kfoundation.org/it-is-no-measure-of-health-to-be-well-adjusted-to-a-profoundly-sick-society/> (accessed 18 March 2025).

4 Evie Sun, 'The impact of technology on our epidemic of loneliness', [no date], available at <https://brainmindsociety.org/posts/the-impact-of-technology-on-our-epidemic-of-loneliness> (accessed 18 March 2025).

According to traditional cultures, mind, body, spirit, community and Country are co-constituted rather than split, meaning that spiritual injury is inseparable from ecological injury. Indigenous scholar Leanne Betasamosake Simpson views land not simply as territory or a resource, but as pedagogy and relation – a living presence through which knowledge, ethics, and identity are continually renewed.[5]

Tyson Yunkaporta similarly emphasises that Aboriginal knowledge systems consider health as intertwined with the collective and ecological.[6] A person's wellbeing, therefore, cannot be separated from the health of kin networks, ecosystems, and Country. Indigenous Australians' experiences of environmental degradation are often accompanied by a deep sense of cultural and spiritual loss, including the grief associated with damaged sacred sites and disrupted cultural practices.[7] For the traditional owners of Country, ecological malaise corresponds with cultural, spiritual, and psychological harm. Healing therefore, requires not just psychological repair, but the renewal and rebuilding of close spiritual ties with mother Earth and community.

For Australian First Nations people, like other Earth-based cultures around the world, the psychologisation of climate destruction makes little sense. It also jars with another body of spiritual thought, 'ecodharma', which refers to the integration of Buddhist teachings on wisdom and compassion with ecological awareness and activism that grounds spiritual practice in the urgent work of responding to our environmental predicament.[8] Such spiritual practice is encouraged in part to address the experience of place-based distress or 'solastalgia' caused by ecological loss and fractured attachments – or, as environmental philosopher, Glenn Albrecht, puts it, 'the homesickness you feel while still at home'. For Albrecht, the suffering created by solastalgia can only be addressed through healing practices that integrate care for body, mind, and spirit. This means creating support systems that blend emotional care, cultural practice, and ceremony on Country, along with collective meaning-making as opposed to merely relying upon symptom

5 Leanne Betasamosake Simpson, *As we have always done: Indigenous freedom through radical resistance* (Minneapolis: University of Minnesota Press, 2017).

6 Tyson Yunkaporta, *Sand talk: how Indigenous thinking can save the world* (Melbourne: Text Publishing Company, 2019).

7 Campbell W. Rigby, Allan Rosen, Helen L. Berry and Clive R. Hart, 'If the land's sick, we're sick: the impact of prolonged drought on the social and emotional well-being of Aboriginal communities in rural New South Wales', *Australian Journal of Rural Health* 19/5 (2011), 249–54.

8 Stephanie Kaza, *Green Buddhism: practice and compassionate action in uncertain times* (Boulder, Colorado: Shambhala Publications, 2019); David R. Loy, *Ecodharma: Buddhist teachings for the ecological crisis* (Boston: Wisdom Publications, 2019).

management. A volunteer firefighter, who is for example experiencing arrhythmias and survivor guilt following a bushfire, might find steadiness and calm through debriefing with peers and mindfulness on Country. Or flood-affected locals could clean houses and then sit in yarning circles after a flood event and share their thoughts and feelings with empathetic others.

Most western societies are manifestly ill-prepared to deal with the anguish and bewilderment that will increasingly pervade our senses. Governments might (and must) invest in more levees, stormwater drains, roads, and bridges, but the real challenge lies in building civic connections and support systems to withstand what lies ahead. We have, as yet, no clear answers as to how to achieve all this. Many non-Indigenous cultures remain anchored in ways of being that leave them unprepared for these emotional and spiritual challenges. They are often bereft or unaware of the deep relation structures that sustain life. Indeed, reducing climate disruption to emissions curves or mortality rates is to avoid the big questions that ultimately, each of us must face. Building solidarities across social divides remains essential if we are to respond collectively and effectively to what lies ahead. As Bobbi shared during our conversation, '*we've always found ways to survive all the changes that have come at us, but we've never done it alone. It's always, always, always in community*'.

Despair: The shadow we must face

Despair is a natural and unavoidable response to the dark knowledge of ecological and social collapse. For many, accepting that climate disruption is not a distant threat but an immanent and alarming reality means that everything has to be rethought. The Great Unravelling constitutes an upending of some of our most embedded assumptions about the world and our place in it. For young people in particular – for it is they who will have to brace for the most catastrophic impacts – the future is becoming ever more uncertain and troubling. One survey of 10,000 young people across 10 countries, for instance, showed that over 45 per cent stated that climate anxiety had negatively affected their lives and functioning, with many describing feelings of helplessness, hopelessness, and betrayal by governments.[9] These feelings sit within a broader existential despair that is reshaping how each of us is experiencing everyday life.

9 Caroline Hickman, Elizabeth Marks, Panu Pihkala, Susan Clayton, Rebecca E. Lewandowski, Emily E. Mayall, Britt Wray, Christopher Mellor and Lise van Susteren, 'Climate anxiety in children and young people and their beliefs about government responses to climate change: a global survey', *The Lancet Planetary Health* 5/12 (2021), e863–e873.

It becomes harder to make plans or to feel confident that our hopes and aspirations can be realised. Assumed life-paths feel far less attainable than they once did. Psychologically, climate despair arises when people front up against overwhelming evidence of threat but see no credible way forward. Studies of ecological grief report that despair is particularly common when individuals feel overwhelmed and powerless in the face of accelerating planetary change[10] – the threats become too much to bear, percolating into every aspect of our daily lives. Neuroscientific studies report that repeated, chronic exposure to uncontrollable stressors tends to activate the same chemical-neural pathways found in experiences of helplessness – a condition marked by a consuming sense of futility which suppresses the desire to act, leading to a kind of emotional stultification and affective inertia.[11] Prolonged exposure can manifest as anxiety, depression and deep purposelessness – among the most debilitating of emotional states.

But despair, for all its consuming power, does not need to be the endpoint. While this sort of emotional response can paralyse, it can also act as a catalyst for change – a signal that old coping strategies no longer serve us. Research on post-traumatic experience indicates that individuals and communities often emerge from crisis with strengthened relationships, renewed priorities, and deeper and more focused sense of meaning and purpose.[12] Likewise, social movement studies suggest that moments of collective despair often precede breakthroughs in solidarity and mobilisation, as individuals move from private grief to public action.[13] Climate scientist Joëlle Gergis argues that the act of facing despair honestly and with courage is the precondition for meaningful engagement with situation we now face. She adds that hope is not something we feel, it is something we create by acting.[14] In the *Work That Reconnects*, eco-philosopher Joanna Macy insists on honouring despair rather than bypassing it. By turning toward our pain for the world, Macy argues, we discover that despair is the 'other face of love'.[15] In this light, despair signals deep attachment to life – and, once shared and metabolised, can be transmuted into agency. Many spiritual traditions echo the power of this transformative process: Buddhist practice, for example, teaches

---

10 Cunsolo and Ellis, 'Ecological grief as a mental health response'.

11 Steven F. Maier and Martin E. P. Seligman, 'Learned helplessness at fifty: insights from neuroscience', *Psychological Review* 123/4 (2016), 349–67.

12 Richard G. Tedeschi and Lawrence G. Calhoun, 'Posttraumatic growth: conceptual foundations and empirical evidence', *Psychological Inquiry* 15/1 (2004), 1–18.

13 Jochen Kleres and Åsa Wettergren, 'Fear, hope, anger, and guilt in climate activism', *Social Movement Studies* 16/5 (2017), 507–19.

14 Gergis, *Humanity's moment*.

15 Macy and Johnstone, *Active hope*.

that clinging to permanence in an ever-changing world leads to suffering, while letting go opens space for reflection and equanimity. This spiritual orientation reminds us that conscious awareness of the transitory nature of life offers us a foundation for peace amid all the turbulence.

Seen in this light, despair is more than a shadow to be avoided; rather, it can be a companion that, when faced with courage and honesty, can deepen our capacity for solidarity, justice, and care. It marks the point where illusions break down and a more authentic engagement with life becomes possible. Despair is thus an understandable response to what is occurring and, paradoxically, a doorway into another way of being – the place where the unravelling and renewal meet; a place where we can choose to live with renewed intention and life-enhancing purpose.

In place of unexamined anger

Experiencing prolonged periods of anger, even rage, is also a normal response to the emergent realities of ecological collapse. The desire to defy or tear down those deemed responsible for our collective pain can be visceral and devastatingly powerful, but if untempered, it can also be harmful to ourselves and others. The problem with unexamined anger is that it can easily consume the very person who holds it, and turn clarity into bitterness, fuelling cycles of conflict and violence rather than laying the foundations for agency and change. Research shows that chronic, unregulated anger is associated with heightened stress reactivity, increased risk of depression and anxiety, and increased interpersonal conflict, undermining both mental health and social cohesion.[16] As political philosopher Hannah Arendt argued decades ago, violence born of unchecked anger may well destroy, yet 'is incapable of creating power'; it serves largely to undermine the capacity to build solidarities capable of achieving change.[17]

Anger of course, is part of the repertoire of hitman emotions. Yet when it is consciously recognised and channelled, when it is seen for what it is – a psychological and physiological reaction to perceived threat, injustice, frustration, or violation of boundaries – it can be a source of resilience and empowerment.

---

16 Howard Kassinove and Raymond C. Tafrate, *Anger management: the complete treatment guidebook for practitioners* (Atascadero, California: Impact Publishers, 2002); Benjamin M. Wilkowski and Michael D. Robinson, 'The cognitive basis of trait anger and reactive aggression: an integrative analysis', *Personality and Social Psychology Review* 12/1 (2008), 3–21; Jerry Suls, Paula Green and Sadia Hillis, 'Emotional reactivity to everyday problems, affective inertia, and neuroticism', *Personality and Social Psychology Bulletin* 29/7 (2013), 898–908.

17 Hannah Arendt, *On violence* (New York: Harcourt, 1970), 123.

As civil rights activist Audre Lorde has argued, anger can be a vital emotional resource if channelled with thoughtfulness and care, helping us to identify the sources of injustice and then acting collectively on the basis of that knowledge.[18]

Social philosopher Myisha Cherry similarly insists on the ethical potential of what she calls 'Lordean rage' – that is, anger that remains tethered to and tempered by love and justice rather than simple vengeance.[19] Former NSW Fire and Rescue Commissioner Greg Mullins describes how responders' anger and grief during repeated climate disasters helped catalyse the establishment of Emergency Leaders for Climate Action, which went on to influence national debate and policy-making.[20] Similarly, the anger that sometimes simmers among emergency community responders is commonly transformed into actions that are of great benefit to traumatised communities, including the saving of lives and provision of desperately needed resources. The challenge, then, is not to suppress anger but to acknowledge and channel it so as to transform its raw energy into actions that build durable power.

Another obvious danger of anger – a trap that we can easily fall into – is that it can make us self-righteous, quick to point fingers at politicians, corporations, fossil fuel executives, and/or the ultra-wealthy. While it is undeniable that some actors carry much more moral responsibility than others for the ecological disaster, we must not lose sight of our own responsibility and, at times, complacent complicity.

We may not all belong to the ultra-wealthy '1%', but many Australians sit within the global top 10 per cent of income levels, rates of consumption, and per-capita emissions far exceeding those of the majority of humanity. On average, Australians emit around 15 tonnes of CO_2 per person each year – about three times the global average of 4–5 tonnes – thereby placing the country firmly within the globe leaders driving the climate predicament.[21] In fact, recent studies show that

---

18 Audre Lorde, 'The uses of anger: women responding to racism', in Cherríe Moraga and Gloria Anzaldúa, eds, *This bridge called my back: writings by radical women of color* (Watertown, Massachusetts: Persephone Press, 1981), 127–33.

19 Myisha Cherry, *The case for rage: why anger is essential to anti-racist struggle* (New York: Oxford University Press, 2021).

20 Greg Mullins, *Firestorm* (Australia: Viking, 2021).

21 World Inequality Lab, *Climate change and the global inequality of carbon emissions, 1990–2020* (WIL Working Paper 2021/23, Paris School of Economics, 2021), <https://wid.world/document/climate-change-and-the-global-inequality-of-carbon-emissions-1990-2020-world-inequality-lab-wp-2021-23>, accessed 12 May 2025; Oxfam International, *Confronting carbon inequality: putting climate justice at the heart of the COVID-19 recovery* (Oxfam International, 2020), <https://policy-practice.oxfam.org/resources/confronting-carbon-inequality-puttin g-climate-justice-at-the-heart-of-the-covid-62>, accessed 12 May 2025; IEA, 2021, Our World in Data, 2023

the wealthiest 10 per cent of the world's population are responsible for roughly half of all global carbon emissions.[22]

Viewed from this vantage point, those suffering the worst climate impacts could easily turn their anger toward us, questioning our own claims to moral innocence. That said, as noted in Chapter 3, capitalism has a way of ensnaring us all in its destructive habits – and ensnared we are. This however, should not invite guilt or paralysis. What is required is heightened awareness of our role in the order of things, which can shift us from blame toward humility and solidarity, and to some extent, shared responsibility – the foundation for more honest and collective forms of response.

Another challenge we are likely to encounter in an increasingly fragmented society is the tendency to blame certain social groups for our problems. Indeed, there are growing signs of racism, scapegoating and political violence across the world, including in Australia. When collective rage lacks outlets for meaningful action, when it attributes blame for social suffering to certain groups, this can lead to bitterness and hatred, and sometimes, destructive cycles of violence, which demagogues readily exploit such anger, directing legitimate grievances toward exclusion and xenophobia – practices that Hannah Arendt described in *The Origins of Totalitarianism*.[23] Without a clear and well-articulated political project, frustration at systemic abuses can quickly morph into disengagement and cynicism, or even mantras such as 'burn it all down'. Unexamined anger thus risks reproducing the very logics of domination and division that have fuelled ecological and social collapse, underlining the urgency of channelling our emotions into more constructive and life-affirming orientations.

The historical record shows that large-scale nonviolent mobilisation – the building of civic formations and progressive social movements – has been far more effective in achieving strategic aims across the twentieth century.[24] Channelling anger into disciplined, nonviolent civil action serves to protect vulnerable communities, sustains broad coalitions, and lowers the risks of backlash.[25] In practical terms, this means building prefigurative alliances whose strategies and tactics reflect the relational, caring, and regenerative worlds they are seeking to co-create.

---

22 Lucas Chancel and Thomas Piketty, *Carbon and inequality: from Kyoto to Paris* (Paris School of Economics, 2015), <http://piketty.pse.ens.fr/files/ChancelPiketty2015.pdf>, accessed 12 May 2025; Oxfam International, *Confronting carbon inequality*.

23 Hannah Arendt, *The origins of totalitarianism* (New York: Harcourt Brace Jovanovich, 1973).

24 Erica Chenoweth and Maria J. Stephan, 'Why civil resistance works: the strategic logic of nonviolent conflict', *International Security* 33/1 (2008), 7–8

25 Chenoweth and Stephan, 'Why civil resistance works', 7–8.

Marshalled well, anger can become clarity-in-motion: a nonviolent foundation for an organised movement oriented toward peace with justice and collective wellbeing.[26]

For many, entering political struggle feels discordant with values of care, humility, or spiritual depth. Yet political engagement is one of the ways those values are expressed at scale. Voting, organising, advocating, and standing in solidarity are not departures from inner work, but extensions of it – ways of aligning outer action with inner being. There are limits to what personal resilience, ethical consumption, or local action can achieve on their own. Some harms are such that they exceed individual or community capacity. Recognising this is not disempowering; it is what directs attention toward collective political pressure as a necessary complement to inner work.

Radical acceptance and courage

How can we make use of the emotions that trouble us rather than be paralysed by them? We argue that 'radical acceptance', a term coined by clinician Marsha Linehan,[27] can offer an effective means of giving direction to our more challenging emotions. From a place of acceptance, we can begin to acknowledge reality as it is, without denial, avoidance, or judgement. Linehan emphasises that suffering intensifies when we resist what is unchangeable: 'Pain plus non-acceptance equals suffering',[28] she argues. In the context of climate chaos, this insight takes on enormous significance. Paralysis often arises from oscillating between catastrophe-awareness and denial. These polarities offer little more than suffering or temporary reprieve. We need something more.

Radical acceptance therefore does not mean resignation or passivity; rather, it breaks with cycles of avoidance and compulsion, creating relieving psychological spaces to respond with enhanced clarity and purpose. By turning toward the reality of ecological breakdown – however terrifying that appears – we can diminish the wasted energy of fighting against what is already unfolding, and open ourselves to constructive, if painful, engagement with the world as it is – right now.

As climate scientist Joëlle Gergis writes in *Humanity's Moment: A Climate Scientist's Case for Hope*, being present to the unfolding devastation of fires, floods, and droughts is not just an academic exercise but an act of grief and love:

---

[26] Chenoweth and Stephan, 'Why civil resistance works', 7–8.

[27] Marsha M. Linehan, *Cognitive-behavioural treatment of borderline personality disorder* (New York: Guilford Press, 1993).

[28] Linehan, *Cognitive-behavioural treatment of borderline personality disorder*, 102.

a refusal to look away from the unravelling of the world as we know it.[29] In *Work that Reconnects*, Joanna Macy argues that rather than avoiding pain, we should be encouraged to acknowledge grief, fear, and anger as natural responses to an unravelling world. To speak the unspeakable. Acceptance, she argues, unfolds in four stages, beginning with gratitude, which grounds us in what is life-giving; then moving to honouring pain, which affirms sorrow and despair as signs of care; then embracing interdependence as the foundation of all life; and finally, committing to actions that foster solidarity and regeneration. Used in activist training and community resilience workshops, this method seeks to transform despair into agency, and isolation into shared purpose and social solidarity.

Many practitioners have over the years sought to integrate the idea of radical acceptance into their work. In the early 1980s clinical psychologist Stephen Hayes developed Acceptance and Commitment Therapy (ACT), a psychotherapeutic approach that blends mindfulness with behavioural and existential insight. Founded on the proposition that facing unchangeable realities frees us to commit to purposeful, values-driven action, ACT is an approach that has benefitted many adherents. Similarly, renowned spiritual teacher Eckhart Tolle maintains that 'conscious awareness' of the present moment offers a portal to inner freedom that is decoupled from the suffering that comes with resistance to 'what is' or 'the isness of things'. Surrendering to the 'now', or simply being present, Tolle argues, opens space for peace and clear-eyed action.[30] Similarly, psychologist and spiritual teacher Tara Brach emphasises that embracing life 'just as it is' opens the heart not only to pain but also to a fearless compassion that both guides and sustains action.[31]

Despair on the other hand, according to existential philosopher Søren Kierkegaard, is a 'sickness unto death': a spiritual malaise that can only be overcome through the courage to face our finitude.[32] French Algerian philosopher Albert Camus further argued that freedom lies in the courageous acceptance of life's absurdity – refusing illusion while affirming meaning through action.[33] Christan philosopher Paul Tillich has referred to this as the 'courage to be', a

---

29 Gergis, *Humanity's moment.*

30 Eckhart Tolle, *The power of now: a guide to spiritual enlightenment* (London: Yellow Kite, 2016).

31 Tolle, *The power of now.*

32 Søren Kierkegaard, *The sickness unto death: a Christian psychological exposition for upbuilding and awakening*, trans. Howard V. Hong and Edna H. Hong (Princeton, New Jersey: Princeton University Press, 1980).

33 Albert Camus, *The myth of Sisyphus*, trans. Justin O'Brien (New York: Vintage International, 1991).

faith-based commitment ('the ground of being') to affirm life despite anxiety, guilt, and the certainty of death.[34] As Joanna Macy puts it, 'the most radical thing any of us can do at this time is to be fully present to what is happening in the world'.[35] If experienced collectively, courage then becomes a bonding social practice rather than an individual endeavour. In this way, radical acceptance permits us to see mortality not as a paralysing threat but as a clarifying reality – the ground upon which courage becomes possible.

This spiritual insight holds enormous potential for developing the calm and clarity each of us will need to undertake the necessary work of relational resilience. In the context of accelerating climate disruption, deep acceptance allows us to grieve loss and injustice without collapsing into despair, and to move forward in solidarity with others who feel much the same. As Brach notes, acceptance opens a gateway to compassion.[36] For communities and activists, radical acceptance does not erase the enormity of the Great Unravelling nor does it obviate the need for continuing political activism. Rather, it enables a steadier presence, where action can arise from a deeper alignment with care, courage, and compassion. In short, radical acceptance is the precondition for action in a time of endings and renewal. It is a response that in some ways resembles the political practice of truth-telling and 'staying-with' the harms and injustices that have contributed to human suffering.

It resonates too with Michel Foucault's idea of *parrhesia* or fearless speech, which advocates the ethical duty of speaking truth even when it is uncomfortable or risky.[37] It is a refusal to be distracted by false claims and misleading narratives, prompting us to face what is happening, naming the systemic drivers of collapse, and acknowledging that no amount of individual 'green consumerism' can substitute for urgent and wholesale change to the current order of things. It further enables the cultivation of solidarities anchored in mutual care, reciprocity, and justice, particularly when institutional abandonment leaves communities to navigate the consequences of disruption on their own.[38]

---

34 Paul Tillich, *The courage to be* (New Haven and London: Yale University Press, 2014).

35 Macy & Johnstone, 2012, p. 28

36 Tara Brach, *Radical acceptance: embracing your life with the heart of a Buddha* (New York: Bantam, 2003), 5.

37 Michel Foucault, *Fearless speech*, ed. Joseph Pearson (Los Angeles: Semiotext(e), 2001).

38 Kyle Powys Whyte, 'Indigenous climate change studies: indigenizing futures, decolonizing the Anthropocene', *English Language Notes* 55/1–2 (2017), 153–62; Pelling, 2011

Beyond hope and despair, toward deep peace

Manifestly, radical acceptance contrasts with the performative optimism' peddled by governments and vested interests whose aim is to obscure the risks we face while outsourcing responsibility to others. The insistence that 'everything will be fine' if we innovate, adapt or simply remain hopeful, even in the face of mounting evidence to the contrary, is a disingenuous discursive strategy designed to both pacify the public and protect existing power structures.[39]

The disregard for the consequences of policy decisions reflects what scholars of environmental justice view as the continuation of the colonial logic of sacrifice, where the lives and lands of marginalised people become expendable in the pursuit of affluent lifestyles elsewhere.[40] In effect, performative optimism allows those insulated by privilege to avoid despair while worsening the burdens faced by those least responsible for current disaster.

As argued earlier, simple binaries of hope or despair, obscure the complexity of emotional life. Hope can foster denial and deferral; despair can immobilise. Our invitation is to step outside this binary. It means resisting the comfort of denial and distraction, choosing instead to engage with reality directly and to call out the underlying systems driving collapse. The IPCC's warning of a 'rapidly closing window'[41] should foster a sober, ethical *presence* to what is already under way. Staying with the predicament means bearing witness to what is occurring and then contemplating how to respond, even though our heads might be in a spin and our hearts breaking. We are, after all, only human.

Core to this approach is the ethic of *deep peace*: a grounded state of being that honours grief and finitude while acting from a place of inner quietude. Hindu spiritual teacher Ramana Maharshi taught that personal liberation, the freedom from suffering, is not attained by striving for imagined futures but through self-inquiry and surrender, and returning to a place of stillness that

---

39 Annelies Kenis and Matthias Lievens, *The limits of the green economy: from re-inventing capitalism to re-politicising the present* (London and New York: Routledge, 2014); Erik Swyngedouw, 'Apocalypse forever? Post-political populism and the spectre of climate change', *Theory, Culture & Society* 27/2–3 (2010), 213–32.

40 Whyte, 'Indigenous climate change studies'; Nixon, 2011

41 IPCC, *Climate change 2022: impacts, adaptation, and vulnerability. Contribution of Working Group II to the Sixth Assessment Report of the Intergovernmental Panel on Climate Change*, eds Hans-Otto Pörtner, Debra C. Roberts, Marina Tignor, Elvira S. Poloczanska, Katja Mintenbeck, Alegría A. Alegría, Melinda Craig, Sarah Langsdorf, Sonja Löschke, Viola Möller, Aderinto Okem and Babatunde Rama (Cambridge: Cambridge University Press, 2022), <https://www.ipcc.ch/report/ar6/wg2/>, accessed 25 November 2025.

enriches human experience.[42] Indian spiritual teacher Nisargadatta Maharaj, also advocates deep self-inquiry as a necessary means of transcending body and mind, and achieving peace and equanimity. 'Peace', Maharaj remarks, 'is your very nature'.[43] These teachings align with Eckhart Tolle's counsel that the present moment, when accepted without resistance, dissolves the compulsions of both despair and illusory hope, giving rise to a profound sense of inner peace in the here and now – the present moment.[44]

Another guide, British philosopher Rupert Spira, draws on the works of Advaita Vedanta, Sufism, Christian mysticism, and modern Western philosophy, arguing that peace is, 'not something to be achieved, but the background of all experience once the mind ceases to divide reality'.[45] For Spira, awareness itself – the open field of experience – is inherently at rest: peace is thus not acquired but recognised as part of what and who we are.[46] In other words, deep peace is already in us, we simply need to pay attention to it for it to emerge. Embodying stillness prompts us to engage in practices of attentive presence: pausing to witness thought without identification, resting in awareness rather than clinging to outcomes, and returning repeatedly to the body, breath, and our own sense of conscious awareness. By affirming our deep interconnections and sense of common purpose, and engaging practices of shared silence, ritual, and embodied care, it is possible to turn individual repair into collective resilience.

All of this could, of course, be dismissed as yet another indulgent appropriation of eastern traditions – a spin-off from the 'spiritual industrial complex'.[47] What we are suggesting here, however, is that spiritual and emotional work is a vital ingredient of what is needed to build resilience within ourselves and with others. It is what will enable us to respond to the challenges ahead with some semblance of peace and equanimity – stillness in action. Without this, we risk the stranglehold of panic and despair. It is from a place of inner peace that we may find the courage, energy, focus and commitment to undertake the work we need to do in the ruins of industrial modernism. Paradoxically, the climate

---

42 Ramana Maharshi, *Talks with Sri Ramana Maharshi*, 2nd edn (Carlsbad, California: Inner Directions, 2000), 15–16.

43 *Nisargadatta Maharaj, I am that, trans. Maurice Frydman (Durham, North Carolina: Acorn Press, 1999)*, 145–146.

44 Tolle, *The power of now*, 33–34.

45 *Rupert Spira, Being aware of being aware (Oxford: Sahaja Publications, 2017)*, 22–23.

46 Spira, *Being aware of being aware*, 35–36.

47 Jonathan P. Herzog, *The spiritual-industrial complex: America's religious battle against communism in the early Cold War* (New York: Oxford University Press, 2011).

disaster acts as a key that unlocks the door to such peace. Or, as permaculturists like to say, 'the problem is the solution'.[48]

Living in a time of endings, as noted in Chapter 1, offers an opportunity to rethink how we want to be, and what truly matters to us. It allows us to reckon with the realisation that what we seek is already present within us. Regular mindfulness practice such as breath work meditation, body scans, or loving-kindness (metta) practices, have been shown to cultivate present-moment awareness, reduce reactivity to stress, and foster greater emotional resilience.[49] Research has also shown that mindfulness-based programmes such as Mindfulness-Based Stress Reduction (MBSR) reduces anxiety, depression, and stress, with improvements lasting well beyond the end of the course.[50] Brain imaging studies indicate that after only eight weeks of training, participants show less reactivity in the amygdala, the brain's stress and threat detector.[51] Experienced meditators also display quieter brain activity in the regions linked to worry and self-criticism, suggesting a more stable state of calm awareness,[52] what Bobbi called in our conversation, '*strategic patience*'.

Physiological measures point in the same direction. A meta-analysis of dozens of trials found that mindfulness lowers stress hormones such as cortisol,

---

48 Bill Mollison, *Permaculture: a designer's manual* (Tyalgum, New South Wales: Tagari Publications, 1988), 15.

49 Jon Kabat-Zinn, *Full catastrophe living: Using the wisdom of your body and mind to face stress, pain, and illness* (New York: Delacorte, 1990); Richard J. Davidson and Alfred W. Kaszniak, 'Conceptual and methodological issues in research on mindfulness and meditation', *American Psychologist* 70/7 (2015), 581–92; J. David Creswell, 'Mindfulness interventions', *Annual Review of Psychology* 68 (2017), 491–516.

50 Madhav Goyal, Sonal Singh, Erica M. S. Sibinga, Nancy F. Gould, Ashley Rowland-Seymour, Ritu Sharma, Zachary Berger, David Sleicher, Daniel D. Maron, Husam M. Shihab, Priyanga D. Ranasinghe, Steven Linn, Somnath Saha, Eric B. Bass and Jennifer A. Haythornthwaite, 'Meditation programs for psychological stress and well-being: a systematic review and meta-analysis', *JAMA Internal Medicine* 174/3 (2014), 357–68; Bassam Khoury, Tania Lecomte, Guillaume Fortin, Monique Massé, Philippe Thérien, Vincent Bouchard, Marie-Andrée Chapleau, Karine Paquin and Stefan G. Hofmann, 'Mindfulness-based therapy: a comprehensive meta-analysis', *Clinical Psychology Review* 33/6 (2013), 763–71.

51 Gaëlle Desbordes, Lobsang T. Negi, Thaddeus W. W. Pace, B. Alan Wallace, Charles L. Raison and Elen N. Schwartz, 'Effects of mindful-attention and compassion meditation training on amygdala response to emotional stimuli in an ordinary, non-meditative state', *Frontiers in Human Neuroscience* 6 (2012), 292.

52 Judson A. Brewer, Patrick D. Worhunsky, Jeremy R. Gray, Yi-Yuan Tang, Jochen Weber and Hedy Kober, 'Meditation experience is associated with differences in default mode network activity and connectivity', *Proceedings of the National Academy of Sciences* 108/50 (2011), 20254–59.

thereby lessening strain on the body's stress systems.[53] One trial observed that an eight-week mindfulness programme was as effective as anti-anxiety medications in reducing diagnosed anxiety disorders.[54] Mindfulness also seems to help open people up to biennial effects of gratitude and connection. Studies have shown that people who practice mindfulness regularly report higher levels of gratitude – one of the most effective ways of enhancing mental health. Even a single, short session of mindful breathing can increase feelings of calm.[55] Another study, this time focused on older adults, found that mindfulness training both reduced loneliness and decreased harmful patterns of gene activity linked to inflammation.[56]

While meditation is an effective way of quelling the mind, many find it too difficult to start with. There are many other ways of gaining calm and peace through practices like crafting, reading a book, surfing a wave, practising yoga, walking in nature, whale watching, artwork and even doing nothing (as in literally, sitting in a chair willingly doing nothing, a non-practice known as 'niksen'[57]). The more we create space in our daily lives to practice these things, the more we can build the personal resilience needed to face our predicament and to nurture feelings of contentment and gratitude.

It is possible therefore, to *both* acknowledge the terrible state of our planet and an uncertain future, *and* to develop inner quietude, *thanks to it*. This contrasts markedly with values propagated by the current system in which being still, slowing down or doing nothing have little or no transactional value. Achieving some degree of inner peace though quietude and stillness is as much a political act as it is a way of improving personal wellbeing in a time of great uncertainty. So inner peace is no mere new age indulgence: it holds great potency as a means of resistance to the dominant order, as well as providing the fortitude necessary for collective action.

---

53 Michael C. Pascoe, David R. Thompson, Zachary M. Jenkins and Chantal F. Ski, 'Mindfulness mediates the physiological markers of stress: systematic review and meta-analysis', *Journal of Psychiatric Research* 95 (2017), 156–78.

54 Elizabeth A. Hoge, Eric Bui, Sarah A. Palitz, Nicole R. Schwarz, Margaret E. Owens, James M. Johnston, Mark H. Pollack, Naomi M. Simon and Scott P. Orr, 'Mindfulness-based stress reduction vs escitalopram for the treatment of adults with anxiety disorders: a randomized clinical trial', *JAMA Psychiatry* 79/1 (2022), 13–21.

55 Jans-Beken & Wong, 2019; Loo et al., 2021

56 J. David Creswell, Michael R. Irwin, Lisa J. Burklund, Matthew D. Lieberman, Jesus M. G. Arevalo, Julienne Ma, Elizabeth C. Breen and Steve W. Cole, 'Mindfulness-based stress reduction training reduces loneliness and pro-inflammatory gene expression in older adults: a small randomized controlled trial', *Brain, Behavior, and Immunity* 26/7 (2012), 1095–1101.

57 Olga Mecking, *Niksen: Embracing the Dutch art of doing nothing* (London: Little, Brown Book Group, 2020).

Inter-being and interdependence

For centuries, dominant Western traditions have considered human experience as separate from the natural world, with land, water, and other species viewed as resources to be exploited. This worldview has underpinned colonialism, dispossession, and environmental destruction, treating the Earth as an object-resource rather than a living community. To live sanely in this time of endings, we require a form of conscious awareness that refuses separations – self from other, human from more-than-human, thought from land, justice from ecology. Renowned Buddhist scholar Thich Nhat Hanh names the immanent connection between all living beings as 'inter-being'. He recognises that 'the cloud and the paper inter-are' and offers ways of seeing that dissolve fantasies of autonomy and isolation while restoring reciprocity as the core of life.[58]

Breathing is a sentient expression of this: we inhale what trees 'exhale', and they 'inhale' what we exhale. Indigenous scholars refer to this nexus as relational life: worlds sustained through interdependent responsibilities, obligations, protocols and consent among peoples and beings.[59] Relationships with land and water are not objectified lifelessly as management or ownership but as living entities embracing responsibility, custodianship and kinship. To live well, in this sense, is to honour reciprocal obligations within a web of relations that includes humans, non-humans, and ancestral beings.[60]

Kyle Powys Whyte, a Potawatomi scholar, explains that the idea of Indigenous sovereignty is not simply about political authority over a territory. Rather, it reflects the obligation of maintaining long-term responsibilities to relatives, both human and other-than-human, across generations.[61] Country is a living, breathing entity that is alive, present, and replete with ancestral stories – stories that reflect the life-affirming nature of Country as well as the dark history of its exploitation and plunder by unwelcome strangers. Climate disruption – a manifestation of this tragic incursion – is the continuation of a long history of colonialism which has fractured and, in some cases, eviscerated a delicate ecological balance that has existed over millennia. The reification of nature, its reduction to simple utility, as Nishnaabeg scholar, Leanne Betasamosake Simpson points out, assumes

58 Thich Nhat Hanh, *Peace is every step: the path of mindfulness in everyday life* (London: Random House, 2010) 95–96.

59 Whyte, 'Indigenous climate change studies'; Kyle Powys Whyte, 'Too late for Indigenous climate justice? ecological and relational tipping points', *Wiley Interdisciplinary Reviews: Climate Change* 11/1 (2020), e603.

60 Whyte, 'Indigenous climate change studies'; Simpson, *As we have always done*.

61 Whyte, 'Indigenous climate change studies'.

a hieratical relationship that claims human dominion over all living things. Instead, Simson speaks of traditional practices of resurgence – everyday acts of care, storytelling, and land-based teaching that rebuild reciprocal relationships with the land and indeed, with all living beings.[62]

Tyson Yunkaporta, an Aboriginal scholar from the Apalech clan, alludes to how Indigenous thinking works through patterns and stories rather than abstract, instrumental categories. In his celebrated book, *Sand Talk: How Indigenous Thinking Can Save the World*, he explains that deep time 'pattern thinking' recognises the interconnectedness of systems, where each part reflects and influences the whole. This contrasts sharply with fragmented Western epistemologies that relegates such complexity to extractivist control over 'resources' in the pursuit of private gain. For Yunkaporta, solutions to ecological collapse will not come by imposing new designs on Country but by aligning social and ecological systems with the patterns already present in Country. In other words, sustainability and regeneration – by words of environmental justice – are not about managing nature but listening to it, learning from it, and designing ecologies in harmony with it.[63]

Robin Wall Kimmerer, a Potawatomi botanist we first encountered in the opening chapter, extends this relational consciousness into the realm of economy. She looks at economics not as a system of transaction, accumulation and profit, but as a practice of gift exchange. In *Braiding Sweetgrass: Indigenous Wisdom, Scientific Knowledge and the Teachings of Plants*, she observes that the natural world offers its abundance – food, medicine, shelter, beauty – as a gift. But every gift carries an obligation: to give back, to care, to maintain the circle of reciprocity. Taking without giving back, Kimmerer argues, is a breach of a caring, nurturing relationship – a form of theft. This commitment to gifting and obligation disrupts the idea of the Earth as a limitless warehouse of resources and situates human presence according to principles of gratitude, responsibility, and importantly, restraint.[64]

This shift in understanding from extractivism to reciprocity has far-reaching implications. Land back movements across North America, co-governance arrangements in places like New Zealand (for example, the Whanganui River's recognition as a legal person), and Indigenous-led Landcare partnerships in Australia all embody forms of reciprocity-in-practice. These initiatives reorient governance away from state-centred 'land management' toward shared responsibility grounded in locality, place-based knowledge and kinship. Land regeneration is thus seen as inseparable from the regeneration of peaceful and just relationships

62 Whyte, *'Indigenous climate change studies'*.

63 Yunkaporta, *Sand talk*.

64 Kimmerer, *Braiding sweetgrass*.

between people and place. Peacebuilding scholar John Paul Lederach, calls for the restoration of the 'moral imagination', as a way of situating ourselves 'in a web of relationships – even with [our] enemies' and building processes that hold complexity without falling back into unhelpful binaries of separateness.[65] Donna Haraway's counsel to 'make kin, not babies' acknowledges the gifts of life, cultivating multispecies alliances and practical response-ability rather than investing in techno-salvation or despair.[66]

Accordingly, inter-being and reciprocity are practical principles for building societies capable of withstanding disruption through care and cooperation rather than conflict and violence. Without acknowledgement of the complexity of living systems, climate 'solutions' become brittle and ultimately unworkable.

In steering land management toward relational ethics, we seek to dismantle extractive logics that have driven ecological collapse. Survival with dignity depends not on fantasies of control but on embracing interdependence as the foundation for transformation. Reciprocity and inter-being remind us that mortality is shared – a collective condition calling us into deeper relationship rather than withdrawal.

What emerges across this journey – from grief, despair, anger, to acceptance and peace, from isolation to reciprocity and inter-being – is recognition that the Great Unravelling is the opportunity to begin another journey.

Conclusion

To contemplate mortality in the Anthropocene – and now, the post-Anthropocene – is to confront the most difficult truths of our age: that the Great Unravelling is not a distant risk but a lived reality, that despair and anger are unavoidable companions, and that denial offers only temporary reprieve. Yet even at the twelfth hour there is the possibility of life-enhancing and, dare we say, even joyous transformation. When grief is honoured, despair becomes a teacher; when anger is acknowledged without being allowed to consume us, it sharpens awareness and being.

Radical acceptance allows us to stop exhausting ourselves in resistance to what cannot be changed and to direct our energies toward what remains possible. Related to such acceptance is a quality of steadiness – a deep peace. It does not

---

65 John Paul Lederach, *The moral imagination: The art and soul of building peace* (Oxford and New York: Oxford University Press, 2005), 50–51, 77–78.

66 Donna J. Haraway, *Staying with the trouble: making kin in the Chthulucene* (Durham and London: Duke University Press, 2016), 4–5, 19–21, 118–120.

erase the pain we might be experiencing but provides the ground upon which courage and clarity can grow. This makes it possible to act without feeling shattered and overwhelmed, to love in the midst of loss, and to discover solidarity with others who are walking the same uncertain path.

The consciousness of inter-being transcends the individual. It reminds us that the embodiment of mortality – the shadow embrace that never leaves us – is shared across great spectrum of relations: human and more-than-human, present and ancestral, local and planetary. To live as well as we can in a time of endings, then, is to inhabit these relationships with care, and profound sense of responsibility.

If the Anthropocene is a time of dying, it can also be a time of pause and renewal, of courage, of solidarity, and of reawakening to the fragile beauty of life itself. For want of a better term, this 'soul work' is not simply about attending to our spiritual and emotional lives – although that, of course, is vital. It is linked to all aspects of adaptation – including the material sort – that are needed to support and protect increasingly vulnerable populations. Attending to our feelings and building new bridges and levees may not be as far apart as we might suppose.

Rebuilding civic culture: Resilience through community

> In significant emergencies and disasters, emergency management personnel do not, and never will, have the capability and capacity to solve the emergency threat for every individual at risk.
>
> —2020 BUSHFIRE ROYAL COMMISSION

> There is no power for change greater than a community discovering what it cares about.
>
> —MARGARET J. WHEATLEY

Introduction

In the previous chapter we explored how inner resources of radical acceptance, peace, presence, and inter-being can steady us in times of uncertainty. Yet clearly, inner practices alone are not enough. To live with dignity and purpose and in an age of turbulence and disruption requires us to consider how, in our daily lives, and in the face of enormous challenges, we relate, act, and build together.

This chapter turns to the social and relational dimensions of resilience. It asks how meaning, resilience and connection can sustain both individual lives and collective futures – offering us, in this era of upheaval, the prospect of some sort of renewal. Beyond survival, our true challenge is to revive civic life and foster place-based networks capable of withstanding what is to come. Our personal and collective consolations may come from the affirmation that life is still worth living, and that we are not alone in the work of care and renewal. This invites a politics grounded in community where care, reciprocity, and mutual aid replace the isolating logics of neoliberal capitalism.

Drawing on social and ecological research, lived experience, and Indigenous knowledges, we explore how communities foster connections; how neighbourly networks and mutual aid become lifelines in moments of chaos; and how custodianship and Caring for Country can renew our bonds with land, waters, and more-than-human kin. Importantly, we recognise that solidarity must be inclusive, meaning we must attend to the multiple inequities and vulnerabilities created by poverty, racism, disability, and gender-based violence.

These are vital considerations. We pitch them as a series of thoughts and ideas that may assist us in responding thoughtfully – rather than simply reacting – to life on a hothouse continent. That said, all that follows is contingent. Nothing about our 'environment' can be assumed. Our landscapes and geographies will be transformed. Caring for Country and regeneration – as broad principles – should continue to hold, yes, but we cannot assume that the land for which we care will remain the same. Some places, for instance, will be submerged under water, or be in heat-blasted, uninhabitable zones. In short, the practices of today may not be entirely applicable tomorrow. We may have to rethink everything as the new reality presents itself. What we offer here in not a formula for survival but rather, tentative signposts toward relational practices that will help us cope with what lies ahead.

Holding the line at Tuntable Falls

On a hot November day in 2019, when dry lightning struck the national park above Protesters Falls and Terania Creek in northern NSW, the image of fire that had been smouldering in the back of people's minds became real on the ridgeline. Over the following weeks, as flames crept towards the forest and communities around Tuntable Falls, the limits of official emergency services became starkly visible. At one point, eleven Rural Fire Service (RFS) appliances were on the property where Virginia Waters ('Gin') lives. Yet it was clear there were neither enough crews nor enough hours in the day to do all that was needed to protect homes, access routes, and the ancient forest she loved.

What unfolded was not a story of 'authorities' coming to the rescue, but of a close collaboration between RFS firefighters and local residents who stepped forward as 'defenders' – Elders and young people from Nimbin, The Channon, Wallace Road, Terania and surrounding communities and farms – arrived with utility vehicles, pumps, and machinery; Gin's own adult children were among those who joined the battlelines. Together and in close collaboration with the firefighters, they cut firebreaks, blacked out hotspots, watched the flight of embers, and patrolled back roads and driveways for weeks after the main front had passed. Community members organised their own fire truck, ran 24-hour patrols for 16 days, then 2–3 patrols up until the beginning of January 2020, and kept watch over smouldering logs long after media attention had moved on.

Gin found herself unexpectedly at the centre of this operation. Asked initially just to '*come out and help*', she became one of the de facto coordinators at the staging ground in her community, working side by side with the RFS communications officers. Each morning and evening she and fellow coordinators joined the briefing with local brigade leaders, then helped translate operational plans into tasks for the defenders: who would cut a new break, who would work on things such as building the crossing at the south end of Tuntable, to ensure an emergency escape route if things went wrong. The line between 'firefighters' and 'community' blurred into a shared fireground.

Gin and the team kept rosters, registered people as they arrived, and insisted that everyone sign back in at the end of their shift. They set up a buddy system so that less experienced locals were always paired with someone who knew the terrain, and tried to hold together work health and safety in a context where urgency made shortcuts tempting. Community vehicles ferried food, fuel, and equipment to appliances across the fireground. When the RFS lacked particular items, such as refillable water canisters, the community effectively handed Gin and the team the cheque book so they could purchase what was needed to keep both crews and defenders supplied.

The work was exhausting and, in retrospect, confronting. Gin later spoke of the anxiety that surfaced once the rains finally came and the patrols could cease: the realisation that she had been asking people, some without formal training, to risk their lives on a dangerous fireground, relying on the informal systems that had been set up and the trust embedded in long-standing relationships. Coming to terms with that responsibility took months. Yet she also recognised that in a context of escalating risk and institutional limits, communities sometimes '*just have to do the work that needs to be done*' for the greater good.

Out of that crucible, the community did not simply 'bounce back'. The experience accelerated a deeper reweaving of civic life. The community went on to formalise refuge areas, establish shared UHF radio channels, create a network of Community Care Teams, contribute to the Nimbin Emergency Radio Network, creating a trusted pathway for verifiable information between residents, police, and the RFS. The bushfire response also laid the groundwork for the Nimbin Community Disaster Plan, which emerged directly from these experiences and formalised the shared systems of communication, roles, and coordination that had been improvised on the fireground. This community-led plan, freely accessible online for other communities to learn from, is a spearheading effort in Australia, paving the way for clearer collaboration and integration of community skills with emergency services. For Gin, the fires were also a tipping point in her own trajectory as a reluctant leader: they strengthened her sense of responsibility for

Country and community, and helped propel her into elected office as a Lismore City councillor.

Gin's story is about more than courage in fire-ravaged landscapes. It shows how consolation can emerge as a civic practice: how meaning becomes anchored in the defence of a beloved forest and the history of activism that protected it; how resilience is built through improvised yet enduring systems of coordination, communication, and care; and how connection is deepened as neighbours, agencies, and more-than-human ecologies are drawn into a shared project of survival. In the age of the Great Unravelling, such collaborations between institutional responders and grassroots networks will be crucial, not as romantic tales of heroism, but as practical demonstrations of how people can hold the line together when existing systems begin to unravel.

The consolation of new possibilities

The question then arises: how do we translate the inner orientations discussed in the previous chapter into everyday life in order to build resilient communities – neighbourly local alliances and networks capable of surviving and thriving in an upended world? And why is this important in terms of how we, as citizens, respond to what is unfolding? This requires something of a balancing act, where we remain mindful of the dispiriting effects of sinking into fatalism, while at the same time not buying into the vacuity of magical thinking. Instead, we seek to tread a path that accepts uncertainty and the unknown and unknowable, while also cultivating ways of being that make life more than bearable.[1]

Consolation in this sense is not about smoothing over pain or offering false reassurances. As Canadian author and former politician Michael Ignatieff puts it, we 'console to find solace together, it is what we do, or try to do, when we share each other's suffering or seek to bear our own. What we are searching for is how to go on, *how to keep going*, how to recover the belief that life is worth living'. To experience and embody consolation, in other words, is to develop the ability to stand with, to accompany suffering without denying it, and to find acceptance, presence and purpose in its wake.[2] Consolation resembles the practice of truth-telling whereby we acknowledge the end of a way of life while

---

1 Macy and Johnstone, *Active hope*.

2 Michael Ignatieff, 'The art of consolation: how to find solace in dark times', *Persuasion* (24 November 2022), <https://www.persuasion.community/p/the-art-of-consolation>, accessed 12 October 2025.

also exploring, affirming, and nurturing what makes us human: our need for meaning, our search for direction, and longing for deep connection.

Consolation is thus experienced not because suffering disappears, but because it is held in relationship and shared purpose. Meaning gives direction, resilience gives endurance, and connection offers care; and community and justice ensure that these gifts are available to all, not only to the privileged or the strong.

When old stories no longer hold

The era of disruption and endings invites us to consider not simply how to survive materially (which we will explore in the next chapter), but how to live meaningfully, especially when old stories no longer hold. Meaning can be derived by investing in something much bigger than ourselves – something that gives us a reason to get up in the morning, and to act. It could be a commitment to life itself, in all its wonder, mystery and complexity. Despite all its travails, life is what we desperately cling to, especially when faced with the inevitability of death. What enriches our lives, however, what makes it more-than-bearable and joyous, even in the darkest hour, is not simply its continuation but the meaning we attach to it. Viktor Frankl, writing out of the horrors of the Holocaust, argued that meaning can sustain life even in instances of extreme adversity, infusing our experience of suffering with purpose and dignity. 'In some way', Frankl remarks, 'suffering ceases to be suffering the moment it finds a meaning, such as the meaning of a sacrifice'. More recent psychological research supports this view, demonstrating that people who can connect their lives to meaningful purpose are more likely to adapt positively to trauma, displacement, and ecological disruption.[3]

For many of us, meaning is found in relationship with family and friends that offers love and care; or in work or activism that contributes to something beyond oneself; or in creative expression, whether music, writing, craft and so forth; or it can be found in spiritual practice or faith traditions that orient life within a larger story; or in connection to place and land; or in rituals and cultural traditions that carry memory and continuity across generations. Meaning then, is not an ethereal spiritual sense, but rather the act of anchoring life in commitments that

---

3 Viktor E. Frankl, *Man's search for meaning* (Boston: Beacon Press, 2006), 104; Crystal L. Park, 'Making sense of the meaning literature: an integrative review of meaning making and its effects on adjustment to stressful life events', *Psychological Bulletin* 136/2 (2010), 257–301; Frank Martela and Michael F. Steger, 'The three meanings of meaning in life: distinguishing coherence, purpose, and significance', *The Journal of Positive Psychology* 11/5 (2016), 531–45.

transcend the individual – caring for others, protecting land, recording events for future generations, capturing the zeitgeist through art, or conveying knowledge across generations. These varied attributions of meaning remind us that even amid chaos and destruction, humans have many ways of anchoring their lives in purpose, continuity, and connection.

In communities, meaningful attachments occur when people share their values through strong relationships, collective storytelling and rituals, and when they act together on things that concern and interest them, reflecting what community psychologists McMillan and Chavis describe as a 'sense of community'. Relationships built on empathy, trust, and mutual care form the social milieu in which individuals feel safe, seen and heard. By sharing challenges and achievements, storying and re-storying, the fruits of collective memory, of our attachments, can be transmitted from one age to another.[4] Meaning within civic communities emerges when individuals recognise a shared purpose or coalesce around a set of values and beliefs. Without this lived sense of the social – of being part of a collective entity – meaning risks becoming private consolation that offers only temporary reprieve. Gin describes this sense of common purpose clearly when acknowledging that when the fires threatened: '*We had a big community response working alongside the RFS…*'. For her, defending the forest and community was '*about the forest as well… that rainforest has never been on fire*' – a reminder of the community's shared responsibilities to place as much as each other.

The bonding that arises on such occasions becomes part of the history of place. It occurs when people actively participate, make joint decisions, and co-create in the context of communal life rather than simply receiving individual support. As Christens argues, empowerment is fundamentally relational, emerging through collaboration, shared influence, and reciprocal recognition – a unified sense of agency that builds resilience and enhances wellbeing and belonging. Ultimately then, meaning is created intentionally through deliberative and inspired actions. It emerges organically through shared experiences, mutual aid, and a common understanding that binds a people together. This has been demonstrated time and again in climate disruption scenarios. The desire to help others, to offer practical and emotional support often overrides the isolation people feel in the normal course of life.[5]

---

4 David W. McMillan and David M. Chavis, 'Sense of community: a definition and theory', *Journal of Community Psychology* 14/1 (1986), 6–23.

5 C Brian D. Christens, 'Toward relational empowerment', *American Journal of Community Psychology* 50/1–2 (2012).

Connection and cooperation of this sort are therefore the stuff of community resilience. The word resilience is often misunderstood as toughness or the capacity to 'bounce back'. Ecological and social research demonstrates that the most effective forms of resilience are relational and systemic: it depends on the diversity of actors, the strength of networks, and the ability to adapt and transform when particular circumstances arise. A community's resilience is enhanced not only by material preparedness – infrastructure, resources etc – but also by trust, skills, cooperation, and inclusive forms of governance. Resilience then, is less about the stoic endurance and more about the collective capacity to improvise, organise and recover, together. This differs sharply from 'prepping' or survivalism, which tends to focus on individual or household self-sufficiency – stockpiling food, weapons, or relying on off-grid technologies in anticipation of societal breakdown. While such orientations may provide short-term security for some, they deepen social separateness by prioritising exclusion, competition, and most worryingly, suspicion of others.[6]

Gin's experiences in the context of terrible, life-threatening fires demonstrates this relational form of resilience. '*Most of the defenders*' she recalls, '*were local people… lots of people and a lot of them were our younger people*'. They worked alongside others as part of an improvised collective. Community members established, '*a line across the community… doing patrols so they can call in anything that was happening*' – a model of shared vigilance rather than isolated self-reliance. Relational resilience therefore, recognises that we are more likely to survive and perhaps live well, when cooperating with one another.

Indeed, very few people can, on their own and all at once, use radio communications, drive a tractor, work among very diverse groups of people, provide medical help, care for the children, grow nutritious food, negotiate their way out of difficult situations, repair a broken generator, use herbal remedies, treat a sick cow, or hold on to their mental resilience. Disaster research consistently shows that the best predictor of survival and recovery is not private stockpiles

---

6 Carl Folke, Stephen R. Carpenter, Brian Walker, Marten Scheffer, F. Stuart Chapin III and Johan Rockström, 'Resilience thinking: integrating resilience, adaptability and transformability', *Ecology and Society* 15/4 (2010), 20; Brian Walker and David Salt, *Resilience practice: building capacity to absorb disturbance and maintain function* (Washington, DC: Island Press, 2012); Daniel P. Aldrich, *Building resilience: social capital in post-disaster recovery* (Chicago: University of Chicago Press, 2012); Fran H. Norris, Susan P. Stevens, Betty Pfefferbaum, Karen F. Wyche and Rose L. Pfefferbaum, 'Community resilience as a metaphor, theory, set of capacities, and strategy for disaster readiness', *American Journal of Community Psychology* 41/1 (2008), 127–50; Ciaran Mills, *Prepared: how to survive in the real world and find the way forward in the crisis* (London: Penguin, 2019).

and gun-toting self-defence but social connectedness and mutual aid.[7] Whereas survivalism tends to reproduce scenarios of scarcity and fear, resilience reframes adaptation as both an individual commitment and collective project built on trust, reciprocity, and the pooling of resources.

Humans are relational beings whose health and survival depend on the nature and quality of their bonds and attachments. This is expressed at three levels. First, as noted in the previous chapter, connection to oneself which involves cultivating self-awareness, emotional regulation (through acceptance and gratitude), and practices that align inner life with outer experience. Research in contemplative psychology shows that mindfulness and self-compassion strengthen wellbeing and adaptive coping mechanisms in the midst of breakdown and turbulence. Second, Robert Putnam demonstrates that social capital predicts not only democratic vitality but also survival during times of great disruption. Interestingly, disaster studies show that mortality rates during heatwaves, earthquakes, and floods are lower in neighbourhoods with strong social ties, even when material resources are similar. This is also true in times when there are no disasters: for instance, in England, the Compassionate Frome Project trained residents as Community Connectors to reduce isolation, leading to a 17 per cent drop in emergency hospital admissions over three years while rates elsewhere rose.[8]

Finally, connection to nature expands the sense of belonging beyond the human. Indigenous traditions attach great value to connection with Country as a living system of law, spirit, and kinship, while ecopsychology has found that time spent in nature fosters resilience, and reduces anxiety and trauma symptoms. These are interrelated aspects of coping in times of disruption, enabling people to feel grounded in self-connection, embedded in caring relationships, and attuned to the more-than-human world.[9]

---

7 Eric Klinenberg, *Heat wave: a social autopsy of disaster in Chicago* (Chicago: University of Chicago Press, 2002); Daniel P. Aldrich and Michelle A. Meyer, 'Social capital and community resilience', *American Behavioral Scientist* 59/2 (2015), 254–69.

8 Jon Kabat-Zinn, *Coming to our senses: healing ourselves and the world through mindfulness* (New York: Hyperion, 2005); Kristin D. Neff, *Self-compassion: the proven power of being kind to yourself* (New York: William Morrow, 2011); Klinenberg, *Heat wave*; Aldrich and Meyer, 'Social capital and community resilience'; Ilona Opper, Mandy Maher, Jane Figgis and Sue Clifford, *Research: social connections – Neighbourhood Connect* (Melbourne: Neighbourhood Connect, 2023), <https://www.neighbourhoodconnect.org.au/research>, accessed 23 October 2025.

9 Yunkaporta, *Sand talk*; Stephen R. Kellert and Edward O. Wilson, eds, *The biophilia hypothesis* (Washington, DC: Island Press, 1993); Martin Jordan, *Nature and therapy: understanding counselling and psychotherapy in outdoor spaces* (London: Routledge, 2009).

Together, these aspects of social and personal fortitude – meaning, resilience, and connection – suggest that human wellbeing is inseparable from the webs of reciprocity we inhabit. This embodied sense of deep connection with community, culture and land is the fundamental architecture of fulfilled human experience that not only embraces history, memories, and identity, but allows us to feel safe and complete. This sense of interconnectedness is thus core to the balance and harmony we need in times of climate disruption and societal breakdown.

An example of this is Resilient Uki (RUKI) in northern NSW, a not-for-profit community group which emerged from mutual aid and community resilience practices forged through repeated disasters. RUKI demonstrates how decentralised street and neighbourhood networks can strengthen our individual and adaptive capacities, while deepening place-based belonging and feelings of attachment. The initiative grew from everyday ties present in community hall activities, community gardens, WhatsApp groups, and tool-sharing – that were already in place, enabling residents to draw on their pre-existing connections to mobilise quickly in response to bushfires, floods, and the pandemic. Over time, this evolved into a flexible yet coordinated approach that collaborated with emergency services while also retaining grassroots autonomy. RUKI illustrates how resilience emerges not from isolation or control but from connection, weaving people, land, and institutions into adaptive, trust-based networks. In Uki, as in countless other communities worldwide, such ties give expression to the assemblages of civic life in practice – the ties of intersectional connection that make purposeful, adaptive solidarities and mutual aid possible.[10]

Community in a time of endings, and beginnings

The word 'community' is frequently trotted out to signify the blue-sky connective moments that arise in both good times and bad. It is, more often than not, imagined rather than real, fleeting rather than enduring. Habitually, when tragedy strikes, we are regaled with stories of how 'the community' has risen to whatever challenge. As sociologists are fond of reminding us, there are countless examples of imagined and actual communities that supposedly reveal the connective tissue that often lies dormant. Yet even though the term has been romanticised and

---

10 Claire McLisky and Emma Pittaway, *Case study: Australia – how mutual aid strengthened Greater Uki's community resilience, 2019–2022* (March 2025), with contributions by Jean Renouf, Jo Longman and Richard Hil, under the direction of Pablo Servigne and Véronique de Geoffroy and Groupe URD, 2025, <https://www.planc.org.au/mutual-aid>, accessed 26 November 2025.

idealised (and sometimes used for exclusionary purposes), its importance in the midst of climate breakdown – as an indefensible means of survival and renewal – cannot be overstated. Without an experiential sense of community – the feeling of being part of a social milieu – it is difficult, especially in the midst of upheaval, to envisage how civic life can function.

If meaning, resilience, and connection offer the inner scaffolding for human fulfilment, then community necessarily provides the architecture through which these qualities are actually experienced. In contrast to the rampant individualism of neoliberal capitalism, it is through lived, felt community that personal sources of meaning become collective practices, where resilience becomes mutual support, and where connection deepens our sense of belonging.

Scholars have long held that community should not be viewed as a fixed, reified entity, nor is it necessarily bound by time and geography. Rather, community is a dynamic, organic process of intersecting relationships and practices which social anthropologist Anthony Cohen sees as the expression of shared meanings, even when people are personally unknown to each other. Similarly, sociologist Amitai Etzioni views community as a balance between individual autonomy and social responsibility, emphasising moral commitments as much as social ties. For British sociologist Gerard Delanty, community is best understood as a series of ongoing communications and negotiations rather than a bounded group – relations that are continually made and remade through interaction. Celebrated sociologist Robert Putnam defines community in terms of social capital, that is to say, the networks of trust and reciprocity that enable and sustain collective action, while social scientist Raven Cretney notes how, in times of crisis, community can be seen as a site of resilience and contestation, where repair, solidarity, and sometimes conflict unfold.[11]

Community therefore is not simply who we like or who lives nearby; it is a set of organic, dynamic and lived practices that foster ties between and among several different individuals and groups. These social intersections shift the idea of community from fluffy, idealised notions to forms of engaged and reciprocal practice, exhibited in systems of cooperation and exchange that are mutually beneficial. Embodied aspects of community are found in tool libraries, seed

---

[11] Anthony P. Cohen, *The symbolic construction of community* (London: Tavistock, 1985); Amitai Etzioni, *The spirit of community: rights, responsibilities and the communitarian agenda* (New York: Crown Publishers, 1993); Gerard Delanty, *Community*, 2nd edn (London: Routledge, 2010); Robert D. Putnam, *Bowling alone: the collapse and revival of American community* (New York: Simon & Schuster, 2000); Raven Cretney, 'Resilience for whom? emerging critical geographies of socio-ecological resilience', *Geography Compass*, 13/9 (2019), e12440.

banks, hyper-local UHF radio communications networks, joint management and maintenance of the local community halls, food exchanges and collective use of community batteries, which enable people to share resources rather than duplicate them. They are also found in initiatives like repair cafés, community gardens and sheds, where people share and learn together, as well as in care activities – such as checking in on one another, cooking for a sick friend, or supporting a family under pressure. These and countless other social activities provide the emotional strength without which civic life could not hold.

While such activities are the glue of social connection, they are somewhat less evident than they once were. One of the great challenges faced by those who advocate the importance of community is that they are pushing against the grain of social isolation and loneliness which, according to some, have reached 'epidemic' levels.[12] The rapid decline of social institutions, from the church to the local pub or club, the ubiquity of online digital communication, and years of forced isolation during Covid-19 have further eroded many people's social skills, confidence levels, ability to empathise, and even the desire to simply reach out to others. What once felt like a natural part of everyday life – the chat over the fence, or unplanned street encounter – can now feel awkward or tiring. We increasingly know less about our neighbours and have few if any friends we can rely upon.[13] We live in a world, as Johan Hari puts it, of 'lost connections'.[14]

There is of course, a political dimension to all this. As political scientists Hannah Arendt warned long ago, isolation and loneliness are not just personal conditions but fertile ground for totalitarianism and ethnonationalism. When individuals are cut off from one another – from the shared spaces of conversation, debate, and collective purpose – they become more susceptible to fear and resentment, less able to discern truth from propaganda, and are easier to manipulate.[15] Rebuilding connection, therefore, is not merely a balm for loneliness but a defence of democracy itself. As noted in the opening chapter, in times

---

12 Richard Hil, Louise Holdsworth and Charlie Brennan, *Being lonely* (Ngara Institute and Resilient Byron, 2020), <https://www.planc.org.au/beinglonely>, accessed 7 April 2025.

13 Melbourne Institute, *The Household, Income and Labour Dynamics in Australia (HILDA) survey: selected findings from waves 1 to 21* (Melbourne: Melbourne Institute, The University of Melbourne, 2025), <https://melbourneinstitute.unimelb.edu.au/hilda>, accessed 23 October 2025; Australian Institute of Health and Welfare (AIHW), *Social cohesion and social connection* (Canberra: AIHW, 2025), <https://www.aihw.gov.au/reports/australias-welfare/social-cohesion-and-social-connection>, accessed 23 October 2025.

14 Johann Hari, *Lost connections: uncovering the real causes of depression – and the unexpected solutions* (London: Bloomsbury, 2018).

15 Arendt, *On violence.*

of intersecting problems, connection becomes the foundation of civic strength: it renews the capacity for solidarity, deliberation, and mutual accountability that prevents despair from turning into submission – or worse, violent hostility. Considered in this light, building community is an act of political resistance, asserting at least the possibility of a shared future against the forces that seek to divide and dominate.

In many ways, therefore, resilience demands that we seek to break with these isolating tendencies, for instance, by joining clubs and associations, becoming active in political movements, limiting screen time, placing greater emphasis on face-to-face engagement, and reaffirming what it means to be caring and empathetic. The act of extending a hand, starting a conversation, or showing up – however daunting that might feel – can be the seed of enduring friendships. These are the everyday bonds that can facilitate the beginnings of meaningful connection and genuine community.[16]

Specific community-building projects can also reap enormous social benefits. For example, Shedding and the Repair Café, both located in Mullumbimby, northern New South Wales, offer opportunities for the learning of practical skills and the development of social capital. Shedding provides a space for people of all ages and backgrounds to come together, learn practical skills, such as making a furniture, leading to improved social connections and improved mental health. The Repair Café offers regular gatherings where volunteers help local people fix broken appliances, clothing, and tools – saving waste from landfill while sparking unexpected conversations and friendships. Participants often remark that what gets repaired is not just an object, but also confidence, trust, and a sense of belonging to place. Similarly, the Library of Stuff, founded in 2016, is a community library also located in Mullumbimby. It is filled with good quality items, from board games to camping gear and gardening tools, that are shared amongst its members, providing access to affordable items for individuals, families and not-for-profit organisations in the Byron Shire. Such places of exchange can turn community from a warm fuzzy feeling into a tangible structure that comes with many social benefits.

The idea of community of course, is not confined to a specific geographical area. It can also be forged around shared interests, identities, and commitments. Digital platforms, professional networks, and mutual aid groups that operate online have become important spaces for solidarity, particularly during crises

16 Hari, *Lost connections*.

like the COVID-19 pandemic.[17] Faith communities, activist collectives, cultural associations, and skill-based guilds can offer the same anchors of meaning, resilience, and connection as neighbourhood groups, while overcoming physical boundaries. The key here is intentional practice: that is, creating rituals of participation, norms of reciprocity, and shared decision-making processes that nurture trust and belonging, whether the 'meeting place' is a street, a Zoom room, or a shared cause.

Drawn from ecological design and permaculture, regenerative thinking also offers powerful means of deepening social connection by modelling human systems on the reciprocity and resilience of natural ecologies.[18] Rather than seeing relationships in transactional or extractive terms, regenerative practices promote cycles of care, reciprocity, and renewal. In practice, this is achieved by creating spaces and places of 'social commons' such as tool libraries, timebanks, or cooperative food systems that mimic ecological cycles of giving and receiving – civic practices that will be desperately needed as societal conditions deteriorate.[19] Permaculture principles – echoed in mantras like 'integrate rather than segregate' and 'use edges and value the marginal' – can be applied socially when bringing diverse groups together, making difference itself a source of resilience.[20]

It is important to emphasise however, that community networks should avoid being co-opted as substitutes for public provision. While mutual aid and community initiatives are vital in times of breakdown, they should not become excuses for states or corporations to offload responsibilities onto already stretched local groups. Social outsourcing of this sort has over recent decades become a common practice among corporations seeking to maximise profits and governments pursuing cuts in public spending.[21]

---

[17] Klinenberg, *Heat wave*; Dean Spade, *Mutual aid: building solidarity during this crisis (and the next)* (London: Verso, 2020); Barry Wellman and Milena Gulia, 'Virtual communities as communities: net surfers don't ride alone', in Barry Wellman, ed., *Networks in the global village: life in contemporary communities* (Boulder: Westview Press, 1999), 331–66; Delanty, *Community*.

[18] Pam Mang and Bill Reed, *Regenerative development & design: a framework for evolving sustainability* (New Jersey: Wiley, 2016); Daniel C. Wahl, *Designing regenerative cultures* (Axminster: Triarchy Press, 2016).

[19] Kate Raworth, *Doughnut economics: seven ways to think like a 21st-century economist* (London: Random House Business, 2017); Kimmerer, *Braiding sweetgrass*.

[20] David Holmgren, *Permaculture: principles and pathways beyond sustainability* (Hepburn: Holmgren Design Services, 2002).

[21] David Featherstone, Anthony Ince, Danny Mackinnon, Kendra Strauss and Andrew Cumbers, 'Progressive localism and the construction of political alternatives', *Transactions of*

Genuine, cooperative resilience however, should ensure that governments and businesses uphold their obligations to provide infrastructure, social safety nets, and adaptive climate action. Communities can lobby for equitable disaster recovery funding, advocate for improved public housing and mental health services, or partner with councils to develop inclusive emergency management plans based on local knowledge and capacity, but they should not carry the burden of resilience alone. Civil society, governments and business sectors in short, need to work alongside each other.

You don't have to be best friends, but are you good neighbours?

Support in times of disruption can often depend less on close friendships than the quality of neighbourly ties. Why? Because the people who live near us are invariably the first to come to the rescue of those affected by disaster. Intimate bonds foster belonging, but it is often 'weak ties' – a familiar stranger, the staff behind the local café's counter, or an online chat forum – that make the difference between safety and danger. Disaster research consistently shows that these forms of everyday connection are critical in times of major breakdown. In Chicago's 1995 heatwave, for example, mortality rates were dramatically lower in neighbourhoods with active street life and local trust, even when socio-economic indicators were otherwise similar. Death rates are significantly lower in neighbourhoods where people know and look out for each other.[22]

In New South Wales, the Community Fire Unit (CFU) programme in which volunteer groups of neighbours trained by Fire and Rescue NSW in collective fire preparedness and response offer a vital line of defence in times of disaster. In addition to providing equipment and basic firefighting skills the programme also serves to strengthen networks of mutual aid within neighbourhoods. Residents report that the process of training together, planning evacuation routes, and sharing responsibility deepens their sense of trust, belonging and readiness. This mirrors findings in other disaster contexts in which collaborative neighbourhood alliances have significantly improved survival and recovery rates. In yet another example of collective fortitude, Resilient Kurilpa (RUK) in Brisbane, operates as a

---

the Institute of British Geographers, 37/2 (2012), 177–82, <http://www.jstor.org/stable/41427938>, accessed 8 August 2025.

22 Mark Granovetter, 'The strength of weak ties', *American journal of sociology*, 78/6 (1973), 1360–80; Klinenberg, *Heat wave*; Aldrich and Meyer, 'Social capital and community resilience'.

voluntary network run by local residents and grassroots organisations, committed to flood resilience. As a localised initiative, RUK develops tools, resources, and coordination for apartment dwellers and floodplain residents, enabling them to prepare and respond, together. Their 'Apartment Toolkit' and neighbourhood planning work illustrate how neighbourly ties – in dense, often anonymised urban settings – can become life-saving networks of care and communication.[23]

Neighbourliness thus does not have to begin with grand projects; it can grow from modest, consistent practices that slowly build trust into the social assemblages of daily life. A simple street contact list can ensure rapid alerts in times of need. A local community noticeboard invites conversation and familiarity, transforming strangers into neighbours. Skill-share Saturdays, where Blue Mountains residents exchange knowledge in gardening, cooking, or even languages, builds both competence and connection. And buddy systems for elderly or isolated residents can help ensure that no one is left alone during extreme weather. These small gestures – random and planned acts of kindness and compassion – can generate the social infrastructure that makes rapid coordination and mutual aid possible in moments of disruption. Just as importantly, they cultivate cultures of attachment and belonging in ordinary times, long before disasters strike, ensuring that resilience is not only reactive but deeply embedded in the everyday.

Across the Northern Rivers of NSW, where both of the authors reside, communities are strengthening local resilience through initiatives such as UHF radio networks for communication during disasters; preparing community halls as equipped disaster response hubs; organising neighbourhoods into local resilience networks with trained street leaders; mapping vulnerable residents and connecting them to tailored support; partnering with local businesses to share skills, tools, and resources; creating shared emergency stores and kits; developing flood- and fire-readiness plans; running regular preparedness and wellbeing workshops; planning for widespread food distribution; coordinating community drills and debriefs; or maintaining active online and social media networks that keep information flowing and neighbours connected before, during, and after crises.

Neighbourliness, however, is not always synonymous with peace and harmony. Communities are sites of difference and tension. Maintaining ties with familiar

---

[23] Eriksen, C. and Gill, N. (2010) 'Bushfire and everyday life: Examining the awareness-action 'gap' in changing rural landscapes', *Geoforum*, 41(5), pp. 814–825.; Jim McLennan and Adrian Birch, 'A potential crisis in wildfire emergency response capability? Australia's volunteer firefighters', *Global environmental change part B: environmental hazards*, 6/2 (2005), 101–7, <https://doi.org/10.1016/j.hazards.2005.10.003>, accessed 8 August 2025; *Resilient Kurilpa* website (no author, no date), <https://www.resilientkurilpa.com/>, accessed 8 August 2025.

strangers is ongoing, hard work, and at times, it can turn sour. What matters is not the absence of conflict but the capacity for repair – the willingness to return, listen, and restore relationships after harm. This can be learned through mutual respect and dialogue, and by drawing upon well-established principles found in nonviolent communication, participatory democracy and sociocracy. Local resilience is thus cocreated through consent-based and non-intrusive modalities, adherence to sound principles and carefully considered processes.[24] Strong communities are built on patience and trust, not simple obligation – although that too is helpful in mutually supportive cultures. The aim should always be solidarity, not policing, and care rather than control.

Importantly, while local ties provide the first line of resilience, they gain strength when linked with broader solidarities such as community resilience networks, climate justice movements, Indigenous alliances, and migrant and diaspora networks which serve to widen horizons and distribute resilience beyond any one place.[25] These alliances connect communities to resources, advocacy, and strategies that no neighbourhood alone can muster. In an era of disruption and breakdown, we need both the intimacy of neighbourly care and the reach of trans-local networks.

An illustration of the latter is the Northern Rivers Community Resilience Alliance (NRCRA) in New South Wales. Formed in March 2024, the Alliance brings together over 60 grassroots, place-based resilience groups across the Northern Rivers – from Grafton to Tweed – to share skills, resources, and coordination. Rather than replacing local action, it strengthens it by providing peer support, shared templates, communication infrastructure, and coordination tools to reduce duplication in emergency response scenarios and to enhance collective capacity. The Alliance also clarifies its role relative to state systems – not substituting for institutional duty, but working to amplify community voices and resilience. In doing so, the Alliance illustrates how local neighbourhood ties, when networked across a wider geographic area, become part of a stronger resilience network that distributes load, knowledge, and mutual aid.[26]

---

24 Norris et al., 'Community resilience as a metaphor'.

25 Whyte, '*Indigenous climate change studies*'; Cretney, 'Resilience for whom?'.

26 Plan C, 'About the Northern Rivers Community Resilience Alliance' (no date), <https://www.planc.org.au/nr-community-resilience-alliance>, accessed 8 August 2025; University Centre for Rural Health, 'Northern Rivers Community Resilience Alliance: a movement for resilience' (2 September 2025), <https://ucrh.edu.au/northern-rivers-community-resilience-alliance-a-movement-for-resilience/>, accessed 8 August 2025.

Creating networks of neighbourly support is therefore less about inventing something new than purposefully reinvesting in practices of mutual care and reciprocity in the circles we already inhabit. Research on community resilience points to some key ingredients involved in building networks of care and cooperation. Shared spaces – whether a community hall, a garden, or a WhatsApp group enable people to gather and coordinate. Shared practices – like disaster planning, the establishment of street coordinators, providing opportunities for locals to upskills and connect with each other by organising a workshops and local events – create rhythms of cooperation that deepen trust and confidence among participants. And shared decision-making – through facilitated conversations, agreed discursive parameters, rotating chairs, and consensus-building – ensure that power and responsibility are distributed rather than concentrated.[27]

Other approaches like asset-based community development encourage a focus on the strengths that already exist in a given locality, while mutual aid networks emphasise community solidarity over the allocation of charity.[28] This may mean beginning with a small, practical step like knocking on a neighbours' door and proposing to organise a neighbourhood meal, mapping local skills and resources, setting up a local online messaging system, or joining an existing initiative.

It is worth pausing to notice whether the ties around you, both neighbourly and communal, are strong enough to hold such expressions of care. Do people nearby know you well enough to check in, and do you feel able to do the same for them? What tangible steps can you take to develop, or nurture such relationship with some of your neighbours and your wider community? Nurturing these everyday relationships – whether across the apartment corridor, through a local group, or within a shared initiative – is less about formal obligation than building trust that can hold in times of stress.

But there are no illusions here. People might not want a hand. Not everyone will want to be part of a larger social entity, and certain personalities and political opinions may not gel. Some entrenched prejudices and dislikes may also prevent neighbourliness. And then there are the separations that are so evident in modern societies, brought on by privatised living and the fetishisation of consumption and

---

27 Aldrich, *Building resilience*; Norris et al., 'Community resilience as a metaphor'; Featherstone et al., 'Progressive localism'; Cretney, 'Resilience for whom?'; Delanty, *Community*; Etzioni, *The spirit of community*.

28 John P. Kretzmann and John L. McKnight, *Building communities from the inside out: a path toward finding and mobilizing a community's assets* (Evanston: Center for Urban Affairs and Policy Research, Northwestern University, 1993); Spade, *Mutual aid: building solidarity during this crisis (and the next)*.

technology. These are formidable barriers which are nonetheless worth trying to breach. Not everyone needs to be on board at all times. Those who are willing and able, however limited in numbers, are the ones to start working with. In due time, nature will remind the others to join.

Addressing our differences and fragilities

To be sure, not everyone is equally positioned to contribute to, or benefit from, community ties and neighbourliness. Poverty and insecure housing, disability and dependence on medical or social services, gender-based violence and unsafe shelter, racism and exclusion from decision-making, or chronic health conditions and inadequate access to care can lead to enormous hardship, social suffering and vulnerability. These inequities can often hinder the creation of cohesive communitas. Disasters and disruptions can further expose and deepen these fissures. As disaster scholars remind us, 'natural disasters' are never purely natural; they are experienced through social structures that determine who has the resources to prepare, evacuate, and recover.[29]

To respond well, communities can embrace what policy theorists refer to as 'targeted universalism', which means pursuing universal goals – safety, health, dignity, belonging – while tailoring policies and strategies to those most at risk.[30] Resilience in this sense depends not on uniformity or blanket responses but on the capacity to hold diversity within the gambit of shared purpose. Political theorist William Connolly refers to this as pluralism: the practice of sustaining life and dignity amid profound difference, where differences coexist without collapsing into either fragmentation or forced consensus.[31] Pluralism recognises that communities are made up of intersecting histories, identities, and fragilities, and that justice requires responsiveness to this complexity rather than its erasure.

In practice then, the appreciation of difference (as opposed to a flatline egalitarianism) means designing systems and practices that protect everyone by meeting the specific needs of the most exposed. Targeted universalism and pluralism together form a moral and practical guide for resilience: one that values difference

---

29 Ben Wisner, Piers Blaikie, Terry Cannon and Ian Davis, *At risk: natural hazards, people's vulnerability and disasters*, 2nd edn (London: Routledge, 2004); Susan L. Cutter, 'The forgotten casualties redux: women, children, and disaster risk', *Global environmental change*, 42 (2016), 117–21.

30 John A. Powell, Stephen Menendian and Wendy Ake, *Targeted universalism: policy & practice* (Berkeley: Haas Institute for a Fair and Inclusive Society, University of California, Berkeley, 2019).

31 William Connolly (2017)

as a source of strength and insists that care be extended to all. Minimally, this must involve vulnerable populations directly in decisions that affect their lives; that is, designing with, not just for, marginalised groups.

For example, disability advocacy organisations, such as People with Disability Australia (PWDA), have sought to co-design emergency management plans with local councils, ensuring accessible communication, transport, and safe shelter.[32] The inclusion of visual and auditory alert systems, accessible evacuation routes, and trained community volunteers who can assist people with mobility or sensory impairments serves to safeguard those most at risk as well as improving safety and coordination for everyone during emergencies.

Other initiatives include organising evacuation transport for those without cars; supporting women's refuges, children and LGBTIQA+ and renters' protections during recovery so that displacement does not become permanent dispossession. Gin's story, at the beginning of this chapter, also reminds us that residents are sometimes hard to reach: after the 2022 flooding events, many roads were cut off due to landslides isolating entire communities, and the RFS '*couldn't get to some areas… it was really dangerous*'. So, Community Care Teams were created to '*go around and do community checks during or after an event*'. This system ensures that people in isolated valleys are not left behind and that '*verifiable information*' can still move between the villages, police, and the RFS, while also '*augmenting*' the emergency services' capabilities.

However, even well-meaning community initiatives can unintentionally reproduce forms of exclusion, for example by relying only on digital communication (excluding those without internet), or by assuming access to private cars during evacuation (disadvantaging low-income households). Co-designed inclusive practices can benefit everyone, but cannot be left to chance. They must be built into decision-making processes from the outset, and continually revised in light of lived experience.

People living in an area impacted by a disaster are affected in a myriad of ways, but women experience the impacts differently. As Associate Professor of social work Wendy Foote and her colleagues recall, women are more vulnerable following disasters and are more likely than men to be supporting children and family members, to experience violence, and have higher mortality and morbidity rates. Gender stereotypes and gender inequality, family and other unpaid care responsibilities, social isolation and economic constraints result in many women being

---

32 Jodie Bailie, Veronica Matthews, Ross Bailie, Michelle Villeneuve and Joanne Longman, 'Exposure to risk and experiences of river flooding for people with disability and carers in rural Australia: a cross-sectional survey', *BMJ open*, 12/8 (2022), e056210-1–e056210-10.

placed in positions of increased vulnerability during an emergency. However, across Australia, it is striking that the vast majority of community resilience leaders, coordinators, and connectors are also women, reflecting long-standing gendered patterns of care and civic labour. Research shows that it is women who consistently take on disproportionate responsibility for organising neighbours, sustaining social networks, and coordinating preparedness and recovery efforts, even though this work is rarely recognised or adequately supported.[33]

As you read all this, we invite you to consider: who, in my community, might be left out of standard preparedness and response plans? What assumptions about ability, housing, income, or identity are shaping my perception of my neighbours? What one step could I take to widen the circle of safety?

Custodianship and Caring for Country

To live well in the Great Unravelling requires not only looking after one another, but also rethinking our relationship with what sustains life – Country itself. Landscapes and geographies will change but ethos of custodianship should inform all our practices. It is here that Indigenous traditions offer profound insights. What matters is not stewardship in the sense of managing nature as an external, reified object, but custodianship as the practice of kinship, where responsibility extends across generations to land, waters, species, and ancestors.

In Aboriginal and Torres Strait Islander worldviews, Country is a living system of law, story, and spirit. It encompasses land, waters, skies, and all beings – human and more-than-human – woven together in complex, reciprocal relationships. To Care for Country is to uphold these relationships through practices such as cultural burning, ceremony, songlines, 'governance', and intergenerational teaching. The upholding of this ancestral practice of custodianship – has great relevance in this time of endings. To be sure, the realities of a hothouse Earth may well render some places near uninhabitable, and the reshaping of geographies may mean that communities will have no choice but to relocate. But where life can

---

33 Wendy Foote, Margaret Alston, David Betts and Tracy McEwan, *Women's leadership and a community 'saving itself': learning from disasters, health and well-being impacts of the Northern Rivers flood 2022* (Open Research Newcastle, 2024), available at <https://hdl.handle.net/1959.13/1493441> (accessed 18 March 2025); Tazrina Chowdhury, Paul Arbon, Mayumi Kako, Robert Muller, Malinda Steenkamp and Kristine Gebbie, 'Understanding the experiences of women in disasters: lessons for emergency management planning', *International journal of disaster risk reduction* 104 (2024), 104244, available at <https://doi.org/10.1016/j.ijdrr.2023.104244> (accessed 18 March 2025).

endure, albeit in an altered state, this is where ancestral knowledges can guide us into regenerative ways of being.[34]

Consider the Indigenous ranger programmes operating across Australia. In Arnhem Land, rangers work with Elders to manage fire through traditional burning practices – 'cool burns' that prevent catastrophic wildfires while also regenerating ecosystems. One ranger describes the work as 'looking after Country so Country can look after us', a simple yet profound articulation of the meaning and gifts of custodianship.[35] Here, ecological restoration is inseparable from cultural continuity, intergenerational knowledge transfer, and sovereignty. For communities immersed in this world of reciprocity, work on and with Country, fosters stronger ties across generations, and illustrates that resilience is about nurturing our 'environment' as much as meeting our own needs.

In the Northern Rivers, the Jagun Alliance brings together First Nations Elders, scientists, and conservationists to protect and regenerate ecosystems through cultural burning and Indigenous land practices. Their work demonstrates that ecological knowledge is inseparable from cultural responsibility, and that genuine custodianship requires centring Indigenous authority in land management. Similarly, First Nations organisations like The Returning and We-Alli embody this living ethic of custodianship in practice. The Returning hosts community gatherings and immersive experiences that reconnect people to place through ceremony, listening, and embodied learning. We-Alli offers a Culturally Informed Trauma Integrated Healing Approach (CITIHA) for individuals, families, communities and organisations, where cultural safety for both Indigenous and non-Indigenous people is viewed as foundational. Grounded in Aboriginal philosophy, *dadirri* – a contemplative practice of deep, attentive listening – it offers a way to hold space for storytelling, healing, and collective learning. Played out in workshops, this approach complements a range of reflective practices such as guided narration, art-making, movement, music, symbolism, ritual, drama, somatic awareness, and emotional release work. While many of these techniques are now reflected in contemporary therapeutic methods, their origins rest in Indigenous knowledge

---

34 Deborah Bird Rose, *Nourishing terrains: Australian Aboriginal views of landscape and wilderness* (Canberra: Australian Heritage Commission, 1996); Samuel Muller, 'Co-motion: making space to Care for Country', *Geoforum*, 54 (2014), 132–41; Jessica K. Weir, Catherine Stacey and Kara Youngetob, *The benefits of traditional fire management practices for biodiversity* (Canberra: CSIRO Ecosystem Sciences, 2011).

35 Weir, Stacey and Youngetob, *The benefits of traditional fire management practices*; Ruth Hill, Fiona J. Walsh, Jocelyn Davies, Annie Sparrow, Moira Mooney, Russell M. Wise and Maria Tengö, 'Knowledge co-production for Indigenous adaptation pathways: transformative learning, action, and renewal', *Ecology and society*, 17/3 (2013), 12.

systems and cultural traditions which, for generations, have enabled Aboriginal peoples to express emotion, resolve conflict, honour Country, and renew relationships with self and community.[36]

But such practices cannot be tokenistic. True custodianship requires long-term relationships, genuine decision-making power, and respect for Indigenous sovereignty. For non-Indigenous people, engaging with Country begins with humility and listening. This means recognising historical and ongoing dispossession, and supporting Indigenous leadership in landcare, conservation, and governance. Communities that embrace custodianship principles must commit to listening, learning, and sharing authority, ensuring that ecological regeneration is not decoupled from the ancestral traditions of those who have maintained these relationships across millennia. Partnerships can be forged that align with Country's patterns, adopting seasonal calendars based on local ecologies, planting native species in community gardens, or collaborating with Traditional Owners on river restoration projects.[37]

As you reflect on custodianship and Caring for Country, consider: what would it mean to treat the places I inhabit not as property to manage or exploit but as a relationship of kin and care? Even as landscapes are altered, perhaps changed for good, how can I engage with them, be with them, nurture and even regenerate them? What small step might I take to walk more gently, and more responsibly, with Country? Are there opportunities in my community to support Indigenous-led initiatives, or to align my daily practices – from the food I eat to the land I walk on – with principles of reciprocity and respect?

Being of service

Caring for Country reminds us that our responsibilities extend beyond the self to land and more-than-human kin, while being of service provides a way to translate that understanding into meaningful, grounded action.

In turbulent times when despair and nihilism tempt withdrawal from community and activism, service can prove a powerful antidote. When life seems uncertain, when the future feels opaque or bleak, service as an ethic of participation

---

36 *We-Alli* website (no author, no date), <https://www.wealli.com.au/>, accessed 8 August 2025; The Returning, 'Our story' (n.d.). Available at: <https://www.thereturning.com.au/our-story>, accessed 23 October 2025.

37 Whyte, *'Indigenous climate change studies'*; Linda Tuhiwai Smith, *Decolonizing methodologies: research and Indigenous peoples*, 2nd edn (London: Zed Books, 2012); (Jackson & Barber, 2013; Ens et al., 2016)

offers orientation: a reason to act, a way to stay human, and a compass that points toward dignity even in the midst of breakdown and chaos. Importantly, service should not be viewed as a form of charity, which can reinforce hierarchies between givers and receivers. Instead, it is better regarded as a practice where we allow something bigger than us to happen through us.

Gin's story speaks directly to this ethic: '*My forest. This is where I want to stay… I feel like I'm owned by it*'. Her commitment to protecting the ancient rainforest during the fires reflects a custodial responsibility shaped by decades of activism: '*our national park wouldn't be there without the people that did the Nightcap and Terania protests in the late Seventies to protect it… they're my heroes*'.

Service becomes sustainable not when it is framed as a laborious duty, but when it is infused with a sense of generosity, joy and conviviality. Shared meals, humour, music, and play can all be seen as aspects of service that bolster rather than deplete ailing spirits. This echoes French authors Pablo Servigne and Raphaël Stevens' notion of '*collapsologie joyeuse*' – the idea of a 'happy collapse', which we mentioned in the opening chapter, where communities find vitality, imagination, and solidarity precisely in the act of facing endings together.[38] In this way, service is understood as a generative practice that nurtures resilience because it nourishes both body and spirit.

The Northern Rivers of New South Wales offers yet other compelling examples of how this works. After the devastating 2022 flooding events, community kitchens set-up by spontaneous volunteers were established across the region. These spaces went beyond providing essential meals. They became sites of joy, laughter, and solidarity, where people cooked together, shared food, stories, played music, and supported one another through grief and uncertainty. In many affected towns, residents described how impromptu community meals and shared cooking in local halls became a way to process trauma, claim a sense of togetherness, and restore belonging. Food became both physical and spiritual nourishment.

Similarly, in Mullumbimby, over 60 volunteers showed up to assist a pro bono counselling service that offered vital help to countless people traumatised by the 2022 floods. Social workers, body workers, mental health nurses, psychologists, psychotherapists, counsellors and even a psychiatrist volunteered their time in what proved to be an exemplary collective expression of altruism and care.

Some of the most effective and nurturing forms of service occur within the rhythms of everyday life – in care circles that coordinate meals, transport, or respite for caregivers, or food cooperatives that share their abundant offerings

---

[38] Servigne and Stevens, *How everything can collapse*.

while at the same time reducing waste. It can also take the form of youth apprenticeships in landcare, ecological restoration, or traditional trades, ensuring that skills and knowledge are passed across generations. Repair cafés and mending guilds provide spaces where not only objects but also relationships are restored, while end-of-life companions – such as death doulas – embody service at life's final threshold, offering presence, dignity, and care when it matters most.

What we are discussing, pleading for, here and throughout this book, is a reconstituted form of active civic life in which each citizen is encouraged to play a role in building social cohesion and cultures of mutual aid. This means too, thinking about our own role in local communities, and what we might do to foster social capital. Small connective gestures can soon morph into deeper attachments that build toward solidarities capable of responding to the challenges before us.

None of this is easy. But to break the mould of social disconnection and therefore immiseration, we must learn to live together again, and take some tiny, tentative steps toward building the collective.

Conclusion

The Great Unravelling strips away illusions of certainty and control, but it does not, and cannot, erase our capacity for connection – something we will need in times of chaos and confusion. How we live, the way we live, and the meaning we derive from these matters a good deal as we seek to co-create new ways of being in an upended world. Relational resilience – a core aspect of active citizenship in the new disorder – offers an antidote to fear and despair. What is needed now is not rugged self-sufficiency of the neoliberal sort, but active recognition of our interdependence and collective capacity to improvise, adapt, and transform.

We speak of such fortitude in the full realisation that the changes wrought by climate disruption and societal collapse will be unpredictable, messy – and that there are no guarantees, least of all of fit-for-purpose communities. Nonetheless, meaningful connection offers the best chance we have for a life worth living. Connection to self, to others, and to nature helps to ground us in feelings of attachment and belongingness. They are the bedrock of that quant term, social capital. The lived sense of community, in the context of the revitalised civic society we urge, is the conduit through which a sense of purposeful solidarity is expressed. Whether via local resilience networks, repair cafés, or neighbourhood street groups, practices of care, deliberation, and reciprocity translate sentiment into vibrant and practice-centred social milieus.

Yet we also recognise that there are enormous social divides, requiring thoughtful, targeted co-designed strategies to ensure that no one is left behind.

Custodianship and Caring for Country remind us too that resilience is inseparable from our relationship with the more-than-human world, and that Indigenous teachings illuminate how respect, reciprocity, and responsibility can be realised in community. Finally, we considered service not as duty but as mutual practice infused with joy, humour, and creativity – an approach that sustains both givers and receivers.

Consolation in a transformed world offers neither the promise of safety or escape. Instead, we recognise that, even in a time of endings, we can live with integrity, generosity, and courage. Collapse, we contend, does not need to mean cruelty and barbarism. Rather, we can each help to prise open a door to renewed sense of being that expresses our deepest, life-affirming shared values.

In the next chapter we shift our attention from social to material preparedness such as food, energy, skills, and infrastructure. But it is worth remembering that these material orientations cannot, indeed should not, stand alone. They must be braided with the emotional-spiritual, social, and ethical aspects explored here.

The way ahead: Exercising agency through collapse and renewal

No matter how bleak or menacing a situation may appear, it does not entirely own us. It can't take away our freedom to respond, our power to take action.

—RYDER CARROLL

Resilience is very different than being numb. Resilience means you experience, you feel, you fail, you hurt. You fall. But, you keep going.

—YASMIN MOGAHED

Act now or swim later

—ANONYMOUS

Introduction

The Great Unravelling is no longer a thought experiment or distant dystopia but the setting in which we now live – one that will define the rest of our lives. Heat, fires, floods, and failing systems are already remaking everyday life across Australia and our region. This chapter begins from that frank appraisal and asks: how do we reduce harm while learning to live as best we can in a hothouse Australia?

To answer this question, we draw attention to a range of practical and ethical considerations that might help support life in a dangerously destabilised climate. We argue that mitigation, adaptation, resilience, and regeneration should together form the backbone of a civic response to disruption and collapse.

Mitigation, primarily through the rapid phase-out of fossil fuels and the reduction of all greenhouse gas emissions, remains essential – not as fantasy of reversal or homeostatic, but as a moral obligation to lessen harm. Adaptation, meanwhile, must be pursued not as a technocratic exercise but as a deeply political, cultural, and relational practice. In recognising the inevitability of cascading and compounding disruptions, we explore how resilience can sustain us through difficult times, and what it takes to shift from merely coping to genuine transformation. Finally, we consider regeneration as a way of redesigning our productive systems and patterns of consumption so as to create equitable abundance rather than depletion.

We embark on this journey through the perspective of justice – acknowledging Indigenous sovereignty and those already on the frontlines of disruption. In an era of rolling disasters, the work ahead is to develop responses that are ethical, just and achievable. Living well amid turbulence therefore, is less about restoring what is lost than preparing for what is coming, and reimagining what may still be possible.

Action as antidote to despair

The first time Dr Rebecca McNaught ('Bec') encountered the gravity of climate science, she walked out of a Melbourne office building and wept. As a young environmental practitioner in the early 2000s, she had joined a Victorian Government programme exploring carbon sinks on private land. Her boss had gathered the team for an intensive two-day session on the work of the Intergovernmental Panel on Climate Change. Somewhere in the pages on ice-sheet melt, she learned that even if all emissions stopped tomorrow, the melting and warming already underway would continue for centuries. '*I felt like I knew a dirty, deep, dark secret*', she later reflected. That moment impacted her '*to the core*', and it set the way for everything that followed.

Not long afterwards, she accepted a volunteer position with the International Federation of the Red Cross in the Pacific, supporting Red Cross societies in Fiji, Tuvalu and Samoa as they considered what climate change meant for their operations. It was here that she encountered the early debates about disaster risk reduction, vulnerability assessments, and community-based resilience – issues she has been '*doing ever since*'. As she puts it, '*my antidote to [despair] is some form of action*'. '*I'll die trying*', she adds, '*I want to be able to look my kids in the eye and say that we tried*'.

Two decades later, living in the Northern Rivers and finishing her PhD, Bec encountered climate disruption at close quarters. The 2022 flooding events were unlike anything she had previously experienced, despite her years working alongside Pacific communities. '*I was gobsmacked*', she said. '*I couldn't believe what I saw, and I couldn't believe what I didn't see*'. The scale of the event, and the limits of formal disaster systems, revealed a gap she had talked about for years – the point at which institutional capacity is overwhelmed and communities have no choice but to rely on themselves.

In the wake of the floods, she joined with neighbours to establish the South Golden Beach, New Brighton and Ocean Shores Community Resilience Team, a network built on linking, leveraging, and connecting people with skills, passions, and local knowledge. *'I'm one grain of sand amongst this incredible movement'*, she emphasises, refusing any hint of personal heroism. Her work is largely relational: bringing community members together; connecting local council, neighbourhood centres and government actors; and strengthening the *'two-way street'* between residents and emergency services.

In the months and years after the disaster, the fledgling group had created a street network, established communication channels, mapped vulnerabilities, and laid the groundwork for what would become a recognised community resilience hub in the Byron Shire. During the 2025 Cyclone Alfred, these formative structures showed their worth in the form of street-level messaging, shared updates on coastal erosion, coordination with council staff, and responsive loops that passed information between households, community networks, and emergency services. Residents later told her, *'We felt connected... we weren't alone this time'.*

The work undertaken as a team also illustrated a deeper truth about collective action: it does not depend on a single skillset, but on the convergence of many. Members of the network include those with backgrounds in wellbeing, community development, radio communications, grant writing, and emergency operations. One member delivered workshops on nervous-system regulation; another coordinated radio testing with neighbouring communities; another shaped the group's Street Connect manual. *'There are so many skills and resources within communities'*, Bec said, *'and you can draw on them'.*

Her own household operates on the same principle. The family keeps a disaster kit, essential documents, and back-up equipment. On occasion, the family has evacuated in anticipation of flooding. With a family member who has additional needs, Bec is clear that capacity fluctuates. *'Sometimes you can lean in, sometimes you can't'*, she says, recognising that resilience depends on capacity at any given time.

Bec's motivation to act is in large part driven by concern for her children, and partly by the sense that doing nothing is not an option. She describes action – even small, imperfect action – as the only way she knows to stay aligned with what she understands about the world. Community meetings, though often taxing, especially in terms of preparation, leave her feeling more connected and grounded. They also provide an anchor for her academic work. Research, she says, only matters if it is consistent with what communities face in practice. *'Otherwise'*, she noted, *'it's not going to pass the pub test'.*

Bec's story reflects the unfolding reality of many hard-pressed and under-resourced communities. Systems often fail under pressure, people step in because they must, and the work of building from the ground up can be slow and exhausting. Energies can soon become depleted, especially when there are repeated threats. Nonetheless, Bec's experience illustrates how hypernormalisation dissolves when mutual practice arises. When systems fail, the fiction of orderly continuity gives way to the reality of mutual dependence, improvisation, and collective care. What appears abstract in policy debates becomes concrete in moments of crisis, revealing both the limits of institutional response and the latent capacities within communities.

Why mitigation still matters

Mitigation refers to the practical and political interventions that slow the climate catastrophe by cutting greenhouse gas emissions or increasing their absorption from the atmosphere.[1] It involves major changes in energy, land use, transport, industry, and consumption to achieve deep cuts to carbon, nitrous oxide, and methane emissions.

In the midst of accelerating breakdown, it may be tempting to declare mitigation lost. Carbon dioxide concentrations keep rising, methane emissions from thawing permafrost and fossil fuel infrastructure are intensifying, and the 1.5 °C threshold is already out of reach. Some therefore argue that attention should now focus solely on adaptation. This, however, is a dangerous illusion. Mitigation still matters – scientifically, ethically, and politically.

Every fraction of a degree avoided reduces suffering.[2] The IPCC has shown that the difference between a 2 °C and 3 °C world is not incremental but catastrophic: hundreds of millions more people exposed to lethal heat, crop failure, coral loss, and higher risks of conflict.[3] Mitigation is therefore not about preserving an illusory safe future but going some way to reducing the human and ecological toll of a deteriorating biosphere.

It is also a matter of justice. Oxfam reports that the wealthiest 10 per cent of humanity are responsible for about half of all emissions, while the poorest already live with the severest impacts.[4] Continuing to extract and burn fossil

---

1 IPCC, *Climate change 2022: impacts, adaptation, and vulnerability.*

2 Gergis, *Humanity's moment*, 134.

3 IPCC, *Climate change 2022: impacts, adaptation, and vulnerability.*

4 Oxfam, *Confronting carbon inequality: putting climate justice at the heart of the COVID-19 recovery* (Oxford: Oxfam International, 2020), <https://policy-practice.oxfam.org/resources/

fuels extends centuries of structural violence rooted in colonialism. Kyle Powys Whyte, argues that climate policy often perpetuates 'colonial continuity' – that is, maintaining extractive relationships with land and Indigenous peoples while disguising them as normal and necessary.[5] To abandon mitigation would be to abandon those already on the frontlines of disaster.

In the Pacific, the moral imperative is clear. Nations such as Vanuatu, Tuvalu, and Fiji – facing existential threat from sea-level rise – have become global leaders in climate diplomacy, calling for a Fossil Fuel Non-Proliferation Treaty. Their message is unequivocal: mitigation is survival.

Yet the term is often riddled with illusions, such as the faith placed in speculative carbon removal and geoengineering. Social scientists Markusson et al. call these 'mitigation deterrence' – promises that future technologies such as direct air capture or bioenergy with carbon capture and storage will save us.[6] Kevin Anderson, a leading climate scientist at the University of Manchester, and Glen Peters, a senior researcher at the Center for International Climate Research, warn that such 'negative emissions' are politically seductive but scientifically risky – in effect, an accounting trick allowing polluters to continue.[7] The assumption that vast future carbon sinks will be available creates what other scholars describe as a 'cultural political economy of delay'.[8] Geoengineering schemes extend this hubris. Proposals to inject reflective aerosols into the stratosphere or fertilise oceans illustrate what philosopher Stephen Gardiner calls the 'moral hazard' of climate policy: once such options are on the table, leaders may lean on them as excuses to further delay real emissions cuts.[9]

---

confronting-carbon-inequality-putting-climate-justice-at-the-heart-of-the-covid-62>, accessed 8 August 2025.

5 Whyte, *'Indigenous climate change studies'.*

6 Nils Markusson, Duncan McLaren, Bronislaw Szerszynski et al., 'Life in the hole: practices and emotions in the cultural political economy of mitigation deterrence', *European journal of futures research*, 10 (2022), article 2, <https://doi.org/10.1186/s40309-021-00186-z>, accessed 8 August 2025.

7 Kevin Anderson and Glen Peters, 'The trouble with negative emissions', *Science*, 354 (2016), 182–83, <https://doi.org/10.1126/science.aah4567>, accessed 8 August 2025.

8 Nils Markusson, Duncan McLaren and David Tyfield, 'Towards a cultural political economy of mitigation deterrence by negative emissions technologies (NETs)', *Global sustainability*, 1/e10 (2018), 1–9, <https://doi.org/10.1017/sus.2018.10>, accessed 8 August 2025.

9 Stephen M. Gardiner, *A perfect moral storm: the ethical tragedy of climate change* (Oxford: Oxford University Press, 2011).

Nature-based solutions like reforestation, wetland restoration, and regenerative agriculture can sequester carbon while restoring ecosystems.[10] Yet ecosystem sinks are finite and fragile: forests burn, peatlands dry, mangroves are cleared. Researchers have also found that relying solely on nature-based solutions would require more land than is available to achieve global net zero emissions.[11] Moreover, large-scale land sequestration can displace Indigenous peoples – what scholars Fairhead, Leach and Scoones term 'green grabbing'.[12] Thus, while the IPCC affirms the importance of nature-based solutions', they must complement, not replace, fossil fuel phase-out.

Genuine mitigation means confronting the political economy of fossil dependence: that is to say, shutting down coal, oil, and gas infrastructure, ending subsidies, and redirecting finance to regenerative industries. In Australia – one of the world's largest fossil fuel exporters – this means halting new projects, supporting regional diversification to help economies transition to a low-carbon future by shifting away from fossil fuels, and strengthening the Safeguard Mechanism by requiring real on-site emissions reductions rather than continuing reliance on offsets and credit-mechanisms.[13] It also means, as noted in Chapter 2, addressing the political influence of fossil fuel lobbyists and sections of the corporate media, and reframing economic success around wellbeing and ecological regeneration rather than growth. It means too, as previously noted, exposing the hypocritical nature of government policies which continue to support dirty energy.

Any process of transition away from fossil fuels needs to be just, which means recognising that decarbonisation will have major impacts on livelihoods, especially for those in resource-dependent regions. Workers and communities must be

---

10 Bronson W. Griscom et al., 'Natural climate solutions', *Proceedings of the National Academy of Sciences*, 114/44 (2017), 11645–50, <https://doi.org/10.1073/pnas.1710465114>, accessed 8 August 2025; Nathalie Seddon, Anne Chausson, Pam Berry, Cecile A. Girardin, Alison Smith and Ben Turner, 'Understanding the value and limits of nature-based solutions to climate change and other global challenges', *Philosophical transactions of the Royal Society B*, 375/1794 (2020), article 20190120, <https://doi.org/10.1098/rstb.2019.0120>, accessed 8 August 2025.

11 Gergis, 'Net zero distracts from scientific imperative to eliminate fossil fuels'.

12 James Fairhead, Melissa Leach and Ian Scoones, 'Green grabbing: a new appropriation of nature?', *Journal of peasant studies*, 39/2 (2012), 237–61, <https://doi.org/10.1080/03066150.2012.671770>, accessed 8 August 2025.

13 Ketan Joshi, 'The safeguard mechanism's pro-fossil flaws – explained' (11 August 2025), <https://australiainstitute.org.au/post/the-safeguard-mechanisms-pro-fossil-flaws-explained>, accessed 8 August 2025.

supported through retraining, public investment, and democratic participation.[14] Without justice, mitigation risks deepening inequalities and discontent. When grounded in equity and inclusion, the shift away from dirty energy can also catalyse renewal – restoring ecosystems, decentralising energy, and redesigning economies around sufficiency and reciprocity.[15] Community-owned renewable energy projects, local food systems, and circular manufacturing – designing for durability and reuse, using recycled and renewable materials, and the feedback of waste materials – illustrate how emissions reduction can variously contribute to both social cohesion and ecological repair.

This can be further enhanced through the choices that people make. Research shows that living car-free, reducing flights, adopting plant-based diets, and having smaller families can cut several tonnes of emissions per person each year.[16] However, while such actions have indeed increased among western and other populations, ecology observer George Monbiot argues that small changes on their own, which in some instances he dismisses as 'micro-consumerists bollocks',[17] are not enough.

Still, such actions could gather pace if they were supported by enabling policy.[18] Nonetheless, there is much that each of us can do to help cut emissions and strengthen the push toward a fairer, low-carbon future. Joining or supporting community renewable energy cooperatives, taking part in local regenerative projects, or backing divestment campaigns can all have a real impact. Citizens can also use their voices and vote to hold leaders accountable – demanding they commit to rapid fossil-fuel phase-out and accelerate investment in clean energy

---

14 Peter Newell and Dustin Mulvaney, 'The political economy of the just transition', *The geographical journal*, 179/2 (2013), 132–40.

15 Raworth, *Doughnut economics*.

16 Seth Wynes and Kimberly A. Nicholas, 'The climate mitigation gap: education and government recommendations miss the most effective individual actions', *Environmental research letters*, 12 (2017), article 074024.

17 George Monbiot, 'Capitalism is killing the planet – it's time to stop buying into our own destruction' (30 October 2021), <https://www.theguardian.com/environment/2021/oct/30/capitalism-is-killing-the-planet-its-time-to-stop-buying-into-our-own-destruction>, accessed 8 August 2025.

18 Diana Ivanova, John Barrett, Dominik Wiedenhofer, Biljana Macura, Max W. Callaghan and Felix Creutzig, 'Quantifying the potential for climate change mitigation of consumption options', *Environmental research letters* (2020), <https://dx.doi.org/10.1088/1748-9326/ab8589>, accessed 8 August 2025; Anders Levermann, 'Individuals can't solve the climate crisis. governments need to step up' (10 July 2019), <https://www.theguardian.com/commentisfree/2019/jul/10/individuals-climate-crisis-government-planet-priority>, accessed 8 August 2025.

and just transition policies. Where citizens choose to spend and invest their money matters too: shifting savings and superannuation investments, or refusing to buy from carbon-intensive companies sends a powerful message that the old model is no longer acceptable. Every one of these actions, when multiplied across communities, helps build the social and political momentum needed for large-scale change. As Leiserowitz et al. observe, when collective norms shift, markets and institutions often follow, although the process can be painfully slow and certainly not significant enough to achieve the rapid and wholesale change we so desperately require.[19]

Mitigation, of course, is not merely about parts per million of carbon dioxide; it necessitates moral clarity and political courage, perhaps echoing Martin Luther King Jr.'s 1967 call for a 'revolution of values'. But as essential as mitigation remains, it is no longer sufficient on its own. Even if emissions ceased today, as noted in Chapter 2, the inertia of the climate system would guarantee decades of rising seas and intensifying heat. The science is unambiguous about this.

Further, and as already mentioned, a combination of factors is working against the urgent cessation of greenhouse gas emissions, not least, the ongoing contradictory policies of supposedly 'green' governments; the Ukraine and Iran wars which have prompted many nations to look for new, scaled-up oil and gas reserves; the emergence of authoritarian states with a penchant for extractivism, and the failure of COP meetings to achieve significant reduction targets. Because the primary drivers of climate disruption are political and economic, mitigation cannot be achieved through individual behaviour change alone. Decisions about fossil-fuel expansion, subsidies, infrastructure investment, and regulatory thresholds are made through electoral processes, legislative frameworks, and contested public debate. Engaging in these arenas – by supporting policies and representatives committed to fossil-fuel phase-down – is therefore an integral part of mitigation, not a separate or secondary concern.

Seen through the lens of hypernormalisation, mitigation is not only a matter of emissions accounting but of narrative disruption. Demanding the rapid phase-out of fossil fuels, opposing new extraction projects, and challenging subsidy regimes are acts that puncture the illusion that incremental adjustment within the existing system is sufficient. Mitigation becomes, in this sense, a refusal to participate in the stories that allow harm to continue under the guise of progress. It therefore

---

19 Anthony Leiserowitz, Edward Maibach, Seth Rosenthal, John Kotcher, Jennifer Carman, Lisa Neyens, Teresa Myers, Matthew Goldberg, Emily Campbell, Karine Lacroix and Jennifer Marlon, *Climate change in the American mind, April 2022* (New Haven, CT: Yale Program on Climate Change Communication, Yale University and George Mason University, 2022).

extends beyond technical solutions into the political arena – where decisions about extraction approvals, subsidies, infrastructure and regulation are made, contested and defended. The main challenge at this critical juncture, is to seek to avoid the worst climate scenarios by holding mitigation and adaptation together in ways that reduce harm and elevate our chances of survival and renewal.

Why adaptation is equally urgent

If mitigation remains essential, adaptation is no less urgent. The IPCC defines adaptation as 'the process of adjustment to actual or expected climate and its effects, in order to moderate harm or exploit beneficial opportunities'.[20] In practice, it encompasses everything from raising seawalls and redesigning infrastructure to enhancing crop production, preparing health systems, and strengthening social care networks. Adaptation is the work of rapidly transforming societies to give people and communities a fighting chance of survival and a life worth living.

Laudable though this might be, the word adaptation has become equated with some ill-conceived response ideas. Why so? Let's first distinguish between what energy policy scholar Benjamin K. Sovacool, refers to as 'hard' and 'soft' adaptation.[21] Hard adaptation relies on technological control and capital-intensive projects such as seawalls, engineered flood barriers, and new irrigation systems. While essential, these can become technocratic and exclusionary, often ignoring deeper questions of justice and power. Soft adaptation, on the other hand, focuses on community resilience, behavioural change, and human rights. It also highlights the danger of what some have called 'adaptation apartheid'[22] – a world in which the wealthy adapt while the poor are left to perish.

In Australia, insurers are already withdrawing from high-risk postcodes, leaving low-income households stranded in floodplains or fire-prone regions.[23] Globally, the pattern is starker still. Pacific Island nations confront the possibility

---

20 Intergovernmental Panel on Climate Change (IPCC), *Climate change 2023: synthesis report. contribution of working groups I, II and III to the sixth assessment report of the Intergovernmental Panel on Climate Change* (Geneva: IPCC, 2023), 7.

21 Benjamin K. Sovacool, 'Hard and soft paths for climate change adaptation', *Climatic change*, 104/3–4 (2011), 381–413, <https://doi.org/10.1007/s10584-010-9984-5>, accessed 8 August 2025.

22 Sovacool, 'Hard and soft paths for climate change adaptation'.

23 Productivity Commission, *Natural disaster funding arrangements, inquiry report no. 74* (Canberra: Productivity Commission, 2014).

of cultural and geographic erasure as rising seas threaten entire territories,[24] while African farmers endure recurrent droughts with limited government support or adaptive finance.[25] Indeed, as the UN Environment Programme's *Adaptation Gap Report* notes, adaptation finance remains grossly inadequate, with adaptation needs in developing countries five to ten times greater than current funds.[26]

Some international funding however, has assisted in developing nature-based adaptation initiatives that have led to improvements in biodiversity and livelihoods.[27] Improvements to wetlands, mangroves, and reefs help buffer coasts against storms; urban trees reduce heat stress; forests help regulate water cycles.[28] In the Pacific, mangrove restoration has been shown to cut cyclone damage while sustaining fisheries.[29] Yet even these 'solutions' – which are often co-opted by governments as low-cost substitutes for systemic reform – have their limitations. Mass planting need space, good governance, and time to mature. Saplings and fragile ecosystems are also vulnerable to climate extremes that produce coral reef bleaching, forest fires, and damage to mangroves from sea-level rise.

---

24 Caren Farbotko and Heather Lazrus, 'The first climate refugees? contesting global narratives of climate change in Tuvalu', *Global environmental change*, 22/2 (2012), 382–90, <https://doi.org/10.1016/j.gloenvcha.2011.11.014>, accessed 8 August 2025.

25 Olivia Serdeczny, Sophia Adams, Flavio Baarsch et al., 'Climate change impacts in sub-Saharan Africa: from physical changes to their social repercussions', *Regional environmental change*, 17 (2017), 1585–1600, <https://doi.org/10.1007/s10113-015-0910-2>, accessed 8 August 2025.

26 UNEP (United Nations Environment Programme), *Adaptation gap report 2024: come hell and high water – as fires & floods hit the poor hardest, it is time for the world to step up adaptation actions* (Nairobi: UNEP, 2024), <https://www.unep.org/resources/adaptation-gap-report-2024>, accessed 10 November 2025; OECD (Organisation for Economic Co-operation and Development), *Climate finance provided and mobilised by developed countries in 2013–2022: climate finance and the USD 100 billion goal* (Paris: OECD Publishing, 2024), <https://doi.org/10.1787/19150727-en>, accessed 10 November 2025; OECD, *Scaling finance and investment for climate adaptation* (Paris: OECD Publishing, 2024), <https://www.oecd.org/en/publications/scaling-finance-and-investment-for-climate-adaptation_cccc8b52-en.html>, accessed 10 November 2025.

27 International Union for Conservation of Nature (IUCN), *Financing ecosystem-based adaptation: trends, barriers, and opportunities* (2024), available at <https://iucn.org/blog/202511/financing-ecosystem-based-adaptation-trends-barriers-and-opportunities> (accessed 18 March 2025).

28 Seddon et al., 'Understanding the value and limits'.

29 Sebastian Hülsen, Laura E. Dee, Christian M. Kropf et al., 'Mangroves and their services are at risk from tropical cyclones and sea level rise under climate change', *Commun earth environ*, 6 (2025), article 262, <https://doi.org/10.1038/s43247-025-02242-z>, accessed 4 October 2025.

Poorly designed measures can also worsen inequalities – what climate adaptation scholars Jon Barnett and Saffron O'Neill call 'maladaptation'.[30] For example, seawalls that protect wealthy districts can increase flood risk elsewhere, while large desalination projects may undermine ecosystems and Indigenous custodianship of rivers. Adaptation unfolds within political and legal frameworks that determine who is protected, who bears risk, and who is left to absorb loss. Questions of insurance access, planned relocation, ecosystem protection, and disability-inclusive responses are shaped not only by technical capacity but by public policy, law, and advocacy. Without sustained political pressure, adaptation risks reproducing existing inequalities rather than reducing them.

Another danger is that adaptation measures can result in the reproduction of colonial logics. Indigenous environmental justice scholar Kyle Powys Whyte, a Potawatomi philosopher at the University of Michigan, argues that adaptation often treats Indigenous knowledge as a resource to be extracted, with sovereignty and the redistribution of power relegated, if at all, to second order concerns.[31] In Australia, 'cultural burning' is increasingly cited as an adaptation technique, though social scientist Christine Eriksen and geographer Don L. Hankins remind us it is part of a broader cosmology of law, care, and belonging – not a simple technique to be appropriated by non-Indigenous society.[32] To take such practices out of context risks perpetuating the very dispossession that climate disruption exacerbates.

Adaptation approaches also suffer from what might be termed temporal bias. Governments often focus on immediate, or short-term and measurable risk-management outcomes – especially when it comes to asset protection – privileging visible infrastructure over deeper social problems.[33]

---

30 Jon Barnett and Saffron O'Neill, 'Maladaptation', *Global environmental change*, 20/2 (2010), 211–13, <https://doi.org/10.1016/j.gloenvcha.2009.11.004>, accessed 4 October 2025.

31 Whyte, *'Indigenous climate change studies'*.

32 Christine Eriksen and Dale L. Hankins, 'The retention, revival, and subjugation of Indigenous fire knowledge through agency fire fighting in eastern Australia and California', *Society & natural resources*, 27/12 (2014), 1288–1303, <https://doi.org/10.1080/08941920.2014.918226>, accessed 4 October 2025.

33 Johanna Nalau, Blair Preston and Melissa Maloney, 'Is adaptation a local responsibility?', *Environmental science & policy*, 48 (2015), 89–98, accessed 4 October 2025; Barnett and O'Neill, 'Maladaptation'; Russel M. Wise, Ioan Fazey, Mark S. Smith, Sally E. Park, Hallie C. Eakin, Ernst A. Van Garderen and Bruce Campbell, 'Reconceptualising adaptation to climate change as part of pathways of change and response', *Global environmental change*, 28 (2014), 325–36, <https://doi.org/10.1016/j.gloenvcha.2013.12.002>, accessed 4 October 2025; Mark Pelling, Karen O'Brien and David Matyas, 'Adaptation and transformation',

Yet another shortcoming in policy terms is that climate migration remains poorly integrated into adaptation planning. A global review by independent researchers Dennis Mombauer, Ann-Christine Link, and human geographer Kees van der Geest found that most national policies frame human mobility as a threat to be controlled rather than a reality to be managed with compassion and dignity.[34] The fact is that large-scale displacement is already underway. The UNHCR reports rapid increases in forced migration driven by overlapping environmental shocks, conflict, and inequality.[35] Millions are already on the move. The IPCC projects that hundreds of millions will face chronic coastal flooding or permanent inundation by mid-century.[36] This will have significant impacts on immigration policies, especially among wealthier, emissions-rich nations.

Planned relocation – the organised movement of people and assets from high-risk to safer areas – is complex, but can offer safety if grounded in justice, consent, and cultural continuity. Environmental researchers Miyuki Hino, Christopher Field, and Katharine Mach argue that just relocation requires inclusive processes that respect culture and place.[37] Poorly planned, however, it risks trauma and dispossession – especially for Indigenous communities whose ties to Country are irreplaceable.[38] In wealthy countries like Australia, relocation challenges entrenched attachments to property, permanence and growth, and raises difficult questions as to who moves, when, and who pays.[39]

The question of who pays for what in the adaptation space is of course enormously challenging for governments and citizens alike, although reallocating subsidies to adaptation and regeneration and ensuring that corporations pay their fair share of taxes would be a good place to start. Vast sums will be required to not only respond to climate-related migration (both domestic and from overseas),

---

Climatic change, 133 (2015), 113–27, <https://doi.org/10.1007/s10584-014-1303-0>, accessed 4 October 2025.

34 Dennis Mombauer, Ann-Christine Link and Kees van der Geest, 'Addressing climate-related human mobility through NDCs and NAPs: state of play, good practices, and the ways forward', *Frontiers in climate*, 5 (14 March 2023), article 1125936, <https://doi.org/10.3389/fclim.2023.1125936>, accessed 4 October 2025.

35 United Nations High Commissioner for Refugees (UNHCR), *No escape: on the frontlines of climate change, conflict and forced displacement* (Geneva: UNHCR, 2024).

36 IPCC, *Climate change 2022: impacts, adaptation, and vulnerability*.

37 Miyuki Hino, Christopher Field and Katharine Mach, 'Managed retreat as a response to natural hazard risk', *Nature climate change*, 7 (2017), 364–70, <https://doi.org/10.1038/nclimate3252>, accessed 4 October 2025.

38 Whyte, '*Indigenous climate change studies*'.

39 A. R. Siders, 'Managed retreat in the United States', *One earth*, 1/2 (2019), 216–25, <https://doi.org/10.1016/j.oneear.2019.09.008>, accessed 4 October 2025.

but also in helping communities to prepare for extreme weather onslaughts. There is little doubt the latter is of enormous value.

Local and household-level adaptation measures can significantly reduce vulnerability.[40] Practical measures might include retrofitting homes for heat resilience, improving insulation and ventilation, and planting shade trees to cool microclimates. Urban greening and water-sensitive design – rain gardens, green roofs, permeable surfaces – reduce flood risk and heat stress while to some extent restoring biodiversity.[41] In rural areas, supporting local food systems and regenerative agriculture can strengthen food security and soil health, creating resilience against drought and supply shocks.[42] Such measures are most effective when they sit alongside active participation in community emergency networks, such as local disaster preparedness groups or radio communication networks which are linked to faster recovery and lower psychological distress after extreme events.[43]

Adaptation then, is a necessary response to a reality already upon us. As Barnett and O'Neill observe, the real question is not whether adaptation will happen, but what kind.[44] Will it entrench inequality, or will it offer protection and resilience? The answer rests not in technology alone but in ethics – in whether adaptation becomes a euphemism for exclusion, or a commitment to dignity and fairness in a time of disruption. Clearly, wealthier people will be better placed in retrofitting their homes. They are also able to sell up and move to more expensive, safer locations – a process that is already well underway in Australia and many other countries.

The main challenge, as we see it, is to reclaim adaptation as a practice anchored in justice, solidarity, and care. When adaptation is framed as individual self-reliance without corresponding institutional responsibility, it risks reproducing the very dynamics of responsibilisation described in Chapter 3 – shifting risk downward while leaving the structures that generate vulnerability intact.

---

40 Yared A. Abebe, Amir Ghorbani, Igor Nikolic, Nikola Manojlovic, Andre Gruhn and Zoran Vojinovic, 'The role of household adaptation measures in reducing vulnerability to flooding: a coupled agent-based and flood modelling approach', *Hydrol. earth syst. sci.*, 24 (2020), 5329–54, <https://doi.org/10.5194/hess-24-5329-2020>, accessed 4 October 2025.

41 Andrew M. Coutts, Elizabeth C. White, Nigel J. Tapper, Jason Beringer and Stephen J. Livesley, 'Temperature and human thermal comfort effects of street trees across three contrasting street canyon environments', *Theoretical and applied climatology*, 124/1–2 (2016), 55–68, <https://doi.org/10.1007/s00704-015-1409-y>, accessed 4 October 2025.

42 Miguel A. Altieri and Clara I. Nicholls, 'The adaptation and mitigation potential of traditional agriculture in a changing climate', *Climatic change*, 140 (2017), 33–45, <https://doi.org/10.1007/s10584-013-0909-y>, accessed 4 October 2025.

43 Aldrich and Meyer, 'Social capital and community resilience'.

44 Barnett and O'Neill, 'Maladaptation'.

This means recognising that not all people and communities can adapt equally, and that adaptation must include measures specifically designed for those most at risk – what policy theorists call 'targeted universalism'.[45] We can start by strengthening relationships and local networks by supporting Indigenous custodianship and community-led initiatives rather than depending only on governments or infrastructure to keep us safe. Adaptation also calls for solidarity that reaches beyond borders: standing with climate migrants, farmers enduring drought, and island nations facing rising seas. This shared concern can shape a new kind of collective security – one built on empathy and action rather than fear and 'othering'. By demanding fair policies, supporting justice-focused organisations, and using our voices to hold power to account, we help make adaptation not just a technical project but one grounded in justice and basic human rights.

Competing frameworks of adaptation

Across the world, climate adaptation has moved from being an environmental concern to a defining question of governance, rights and justice. Nations, communities, and individuals are experimenting with different approaches to living in the reality of an unstable climate, reflecting more considered views about responsibility, risk, and resilience. Here, we explore three key frameworks shaping the current conversation around adaptation: Australia's national adaptation planning, Jem Bendell's *Deep Adaptation* movement, and the perspectives emerging from the Global South and Indigenous knowledge systems which together, reveal the tensions and possibilities evident in this space.

The national plan

In 2023, the Australian Government released its *National Adaptation Plan* (NAP), the country's first whole-of-government attempt to chart a coordinated path through escalating climate risks.[46] Supported by the *National Climate Risk Assessment*, the plan identifies eight priority areas – critical infrastructure, water, ecosystems, agriculture, health, energy, coasts, and communities – and commits to developing a National Climate Risk Assessment and new climate services to guide policy and investment. On the surface, this approach represents a long-awaited recognition that climate disruption is no longer peripheral to governance, but

---

45 Powell, Menendian and Ake, *Targeted universalism*.

46 Department of Climate Change, Energy, the Environment and Water, *National adaptation plan* (Canberra: Department of Climate Change, Energy, the Environment and Water, 2025).

rather a central concern. The plan aligns Australia with the *Global Goal on Adaptation* established under the Paris Agreement and echoes the latest *IPCC Sixth Assessment Report*, which calls for urgent and targeted realignments of adaptation strategies.[47]

One of the NAP's strengths lies in its institutional clarity. It embeds adaptation in planning systems, links climate services to public decision-making, and acknowledges the pervasive nature of risk across the country. It also states plainly that 'it is prudent to plan for global warming levels of 2°C to 3°C by the end of this century, with temperatures in Australia likely to track higher than the global average'.[48]

Yet its weaknesses are significant, and illustrative of government priorities. Climate adaptation scientists Johanna Nalau and Mark Howden argue, Australia's new National Adaptation Plan, while a welcome step, does not match the scale or speed of disruption already unfolding. They note that many of its proposed actions lack clear pathways for implementation and that the absence of early monitoring mechanisms means we may not know whether interventions are effective until it is too late. Similarly, the Environmental Institute of Australia and New Zealand warns that the proposed adaptation plan is underfunded and overly technocratic.[49] Further, despite rhetorical commitments to feel-good 'resilient communities', it largely imagines adaptation as a service delivered to citizens, not as a collective process of transformation from below. The policy thus ignores the many challenges associated with creating resilient communities. As Bec observed during our conversation, '*Communities don't sit neatly in the boxes government plans assume. They're complex, messy, but also full of different capacities*'.

Australia's national framework, for all its progress, offers a limited understanding of the complexities of adaptation, while continuing to operate within economic and political parameters ill-suited to a world of cascading disasters. The plan's language of resilience and risk management remains tied to what Siri Eriksen, Andrea Nightingale and Hailie call the 'technocratic rationalities'

---

47 UNFCCC, *The Paris agreement* (2015; FCCC/CP/2015/10/Add.1); IPCC, *Sixth Assessment Report* (no date), <https://www.ipcc.ch/assessment-report/ar6/>, accessed 4 October 2025.

48 Department of Climate Change, Energy, the Environment and Water (DCCEEW), *National adaptation plan* (Canberra: Commonwealth of Australia, 2025).

49 Johanna Nalau and Mark Howden, 'Climate change is causing ever more disruption: can Australia's new adaptation plan help?', *The Conversation*, 15 September 2025, available at <https://theconversation.com/climate-change-is-causing-ever-more-disruption-can-australias-new-adaptation-plan-help-265276>,accessed 4 October 2025; Environment Institute of Australia and New Zealand (EIANZ), *National adaptation plan issues paper submission* (no date), <https://www.eianz.org/document/item/7607>, accessed 4 October 2025.

of neoliberal governance, where adaptation is framed as an instrument for safeguarding growth and competitiveness.[50] It imagines that prosperity can be maintained through efficiency and innovation rather than redistribution or systemic change, echoing what political science professor Thea Riofrancos describes as 'climate-smart capitalism'.[51] This sort of instrumental logic, however, is increasingly untenable in an age of intersecting disruptions where discussion/action on the 'big picture' should prevail over growth maintenance and linear management.

Geographers Pelling, O'Brien and Matyas make the obvious point that most national strategies aim to preserve the existing order rather than transform it, while sociologist Amanda Bertana and colleagues warn that being attached to the status quo risks reproducing entrenched inequities by locking adaptation into the very structures that created the problem in the first place.[52] Adaptation scientists Johanna Nalau, Benjamin Preston and Megan Maloney find that devolving responsibility to local actors without providing the level of resourcing required embeds disadvantages and short-termism, mirroring what sustainability economist Russ Wise et al. describe as the 'path dependency' of incremental adaptation.[53] In other words, systems tend to follow familiar trajectories even when those pathways are no longer adequate for the scale of disruption. Beneath its managerial veneer, the NAP reflects what anthropologist Jason Hickel and ecological economist and political ecologist Giorgios Kallis identify as the deeper contradiction of our time: the illusory belief that endless economic growth can coexist within planetary limits which have already been vastly exceeded.[54] True adaptive resilience, therefore, requires more than technical adjustments. It necessitates a new imaginary – that is, the political willingness to rethink economic priorities, and an acceptance that genuine adaptation cannot flourish in a growth-dependent world.

---

50 Siri H. Eriksen, Andrea J. Nightingale and Hallie Eakin, 'Reframing adaptation: the political nature of climate change adaptation', *Global environmental change*, 35 (2015), 523–33, <https://doi.org/10.1016/j.gloenvcha.2015.09.014>, accessed 4 October 2025.

51 Thea Riofrancos, 'Seize and resist: the global supply chain is up for grabs', *The baffler*, no. 54 (November 2020), <https://thebaffler.com/salvos/seize-and-resist-riofrancos>, accessed 4 October 2025.

52 Pelling, O'Brien and Matyas, 'Adaptation and transformation'; Amanda Bertana, Brett Clark, Tabitha Benney and Cameron Quackenbush, 'Beyond maladaptation: structural barriers to successful adaptation', *Environmental sociology*, 8 (2022), 1–11, <https://doi.org/10.1080/23251042.2022.2068224>, accessed 4 October 2025.

53 Nalau, Preston and Maloney, 'Is adaptation a local responsibility?'; Wise et al., 'Reconceptualising adaptation'.

54 Jason Hickel and Giorgos Kallis, 'Is green growth possible?', *New political economy*, 25/4 (2020), 469–86, <https://doi.org/10.1080/13563467.2019.1598964>, accessed 4 October 2025.

Deep adaptation

While governments craft plans, others have paid more attention to how 'ordinary people' confront the psychological and existential aspects of adaptation and build civic solidarities. Among the most controversial in this space is Jem Bendell's *Deep Adaptation*, first self-published in 2018.[55] Bendell argued that societal collapse was not only possible but probable within the lifetime of those alive today. His original framework – the four 'Rs' of resilience, relinquishment, restoration, and reconciliation – invite people to face collapse honestly: to preserve what we most value, to let go of destructive systems, to revive meaningful traditions, and to make peace with loss.

Bendell's work struck a chord with tens of thousands of concerned citizens who felt disillusioned with the faux optimism of mainstream sustainability discourse – which felt terrifyingly at odds with what they were witnessing. It gave permission to grieve, to speak openly about endings, and to form communities grounded in emotional honesty and mutual care. The *Deep Adaptation* network that followed created deliberative spaces for what Bendell called 'collapse literacy' – that is, the shared commitment to nurturing psychological and social capacities to live meaningfully in the midst breakdown. It was a humanising turn in climate discourse, which allowed, indeed encouraged, people to inhabit their despair without being paralysed by it.

Yet Bendell's influence is deeply contested. Critics argue that his framing of the inevitability of societal collapse risks fatalism. John Foster, a philosopher of environmental thought, contends in his 2021 review that the broad church of 'inevitable collapse' in Deep Adaptation pulls in divergent directions and lacks strategic coherence. Chris Saltmarsh, a UK climate justice organiser and writer, warned in 2022 that Bendell's emphasis on collapse and 'acceptance' can divert attention away from political and structural transformation and slip into anti-humanist quietism. According to this reading, acceptance and grief become not a portal to solidarity but a retreat from struggle.[56] Pelling distinguishes between resistance, incremental change, and transformation – asking whether adaptation, deep or light, will simply reinforce the status quo and the structures

---

[55] Jem Bendell, *Deep adaptation: a map for navigating climate tragedy* (University of Cumbria, 2018), 1–31.

[56] Bendell, *Deep adaptation*; John Foster, 'Facing up – facing out or facing in? a review of *Deep adaptation: navigating the realities of climate chaos*, edited by Bendell, J. & Read, R'. (5 July 2021), *Global Discourse Blog*, <https://globaldiscourseblog.co.uk/2021/07/05/facing-up-facing-out-or-facing-in/>, accessed 20 August 2025; Chris Saltmarsh, 'Deep adaptation – or climate justice?' (1 February 2022), *The Ecologist*, <https://theecologist.org/2022/feb/01/deep-adaptation-or-climate-justice>, accessed 20 August 2025.

that underpin power and privilege. O'Brien goes further, describing 'deliberate transformation' as a collective act of agency, a shift in meaning-making as much as political action.

At the same time, more sympathetic scholars have noted that Deep Adaptation has opened up necessary conversations around emotional labour and the limits of hope. Psychologists such as Panu Pihkala, a Finnish scholar of eco-emotions (2022), and Paul Hoggett, a British psycho-social theorist (2021), argue that acknowledging eco-anxiety, mourning, and loss is an essential precondition for meaningful political engagement.[57] Bendell's work then, occupies an uneasy space: it speaks to the affective realities of the Anthropocene while instigating debate about whether emotional authenticity can coexist alongside political struggle. To speak of resilience and adaptation without acknowledging loss is, for Bendell and others, to turn away from truth. The world we inhabit is already profoundly altered: glaciers are retreating, species vanishing, and once-familiar seasons are becoming unpredictable. In short, climate disruption is an unfolding, terrifying reality. As Jem Bendell and geographer Lesley Head both argue, accepting this reality is not surrender but the foundation of psychological and social integrity[58] – it acknowledges what is actually occurring and how people, communities and powerful institutions might respond.

Recognising the limits of human technological know-how and ingenuity is part of this acceptance. Indeed, international climate negotiations reflect a growing recognition that some impacts are beyond adaptation, and that some places will become, are becoming, uninhabitable.[59] Increasingly, communities must navigate 'ambiguous losses', mourning futures that will never come, while finding meaning in what remains.

Australian social work scholars Susan Bailey and Nicholas Gerrish go further, arguing that confronting ecological grief directly can become a basis for radical change in how communities organise care, resistance and adaptation.[60] This tension – between acceptance and agency – marks a deep fault line in contemporary adaptation discourse. Bendell's invitation to honesty and grief resonates

---

57 Pihkala, 'Anxiety and the ecological crisis'.

58 Bendell, *Deep adaptation*; Lesley Head, *Kindred by nature: grief and courage in a climate-changed world* (Sydney: UNSW Press, 2022).

59 UNFCCC, *Report of the Conference of the Parties on its twenty-seventh session: establishment of the loss and damage fund* (Bonn: United Nations Framework Convention on Climate Change, 2023).

60 Sophie Bailey and Nicola Gerrish, 'Ecological grief: how can we bear this together?', *Critical and radical social work*, 13/3 (2025), 348–69, <https://doi.org/10.1332/20498608Y2024D000000059>, accessed 21 November 2025.

precisely because it contrasts with the evasions of political elites. Adaptation is deeply personal and demands clear political direction, the creation of a truly civic solidarity, and structural change. The challenge, then, is to hold the courage to face collapse without surrendering the struggle for justice.

Adaptation as ecological democracy

Beyond the Anglosphere, adaptation takes on other meanings entirely. In much of the Global South, it is not a distant challenge but a lived, everyday reality. Bangladeshi-British scientist and adaptation pioneer Saleemul Huq has long championed community-based adaptation (CBA) – for him, defined as community-driven processes tailored to local capacities, priorities and knowledge. Similarly, Indian environmentalist Sunita Narain argues that adaptation must be built at the point where development and environmental justice converge: policies that link the provision of resources, equitable growth and community agency. In the Pacific – already severely impacted by climate disasters – adaptation often takes the form of what Kiribati leaders refer to as 'migration with dignity':[61] planned relocation that protects cultural continuity, recognises sovereignty and identity rather than treating displacement as rescue or failure – even though it is a tragedy, caused mainly by rich western nations. Meanwhile, recent global tracking shows that only around 17 per cent of adaptation finance between 2017 and 2021 reached the local level, suggesting a persistent disconnect between global commitments and community-level needs.[62]

For many First Nations peoples in Australia – at least for those who can, for now, remain on their lands – Caring for Country is not a climate strategy but a sacred obligation. In North America, Robin Wall Kimmerer speaks of returning to 'the grammar of animacy' – a language that interconnects kinship with the

---

[61] Akka Rimon and Anote Tong, 'The seas are coming for us in Kiribati. will Australia rehome us?' (23 November 2021), *The Conversation*, <https://theconversation.com/the-seas-are-coming-for-us-in-kiribati-will-australia-rehome-us-172137/>, accessed 4 October 2025.

[62] Saleemul Huq, Hannah Reid, Mozaharul Alam, Rachel Berger, Terry Cannon and Andrew Milligan, *Community-based adaptation to climate change: an overview (PLA 60), Participatory learning and action*, 60 (2009), <https://www.iied.org/g02608>, accessed 4 October 2025; Saleemul Huq, 'Improving information for community-based adaptation' (2011), IIED Opinion Paper, <https://www.iied.org/17103iied>, accessed 4 October 2025; Sunita Narain, *Conflicts of interest: my journey through India's green movement* (Penguin India, 2019), <https://www.penguin.co.in/book/conflicts-of-interest/>, accessed 4 October 2025; IIED, 'Shifting climate adaptation finance to local communities: evidence and lessons' (2025), Briefing Paper, International Institute for Environment and Development, London, <https://www.iied.org/22654iied>, accessed 22 October 2025.

entirety of the living world.[63] Kyle Powis Whyte reminds us that Indigenous climate adaptation cannot be separated from the question of sovereignty: it must begin with self-determination, not incorporation into settler colonial frameworks that have, over time, caused so much damage to people and environments.[64]

It is clear from the approaches to adaptation discussed above – from bureaucratic planning to grassroots adaptation, from collapse awareness to Indigenous resurgence – that adaptation is not simply a technical exercise but reflects a range of moral and cultural considerations. It demands new relationships between governments and citizens, and humans and ecosystems. As the IPCC notes, transformative adaptation occurs when governance systems embrace plural knowledges and act upon principles of equity, care and solidarity.[65] It means, in short, listening to the voices of those most affected by climate disruption; those most ignored by wealthy elites – the unpeople, the marginalised and powerless. This should entail the sharing and decentralising of power as much as building responsive capacity. It means too, that instead of fortifying the old, self-destructive world of colonial industrial modernism, adaptation should hospice much of the past and help foster new, decolonised ways of being.

Around the world, communities are already practising forms of adaptation anchored in place, culture, and collective care. In Aotearoa New Zealand, Māori-led restoration projects are rehabilitating wetlands and estuaries, drawing on mātauranga Māori and co-governance models to strengthen ecosystems and cultural identity. In Vanuatu, community-based marine conservation initiatives have integrated customary governance with scientific monitoring, enabling villages to manage nearshore fisheries and protect coral reefs while reinforcing local autonomy. In Latin America, agroecologists such as Miguel Altieri and Clara Nicholls have shown how agroecology – which is the application of ecological principles to the design and management of sustainable food systems – strengthens 'territorial resilience' by restoring soil fertility, biodiversity, and social cohesion simultaneously. These initiatives demonstrate that adaptation can support socio-ecological renewal rather than simply lessening impacts.[66]

---

63 Kimmerer, *Braiding sweetgrass*.

64 Whyte, 'Indigenous climate change studies'.

65 IPCC, *Climate change 2022*.

66 Rachael MacKintosh, 'A climate adaptation model for Māori groups, Aotearoa New Zealand', *Te kaharoa*, 18/1 (2025), 45–68, <https://doi.org/10.24135/tekaharoa.v18i1.483>, accessed 4 October 2025; Ariadne Pascht, Pierre Gubry and Cecilia Anckar, 'Marine conservation in Vanuatu: local conceptualisation and "assemblage"', *Ambio*, 51 (2022), 2389–2400, <https://doi.org/10.1007/s13280-022-01767-3>, accessed 4 October 2025; Miguel A. Altieri and Clara I. Nicholls, 'Agroecology: a brief account of its origins and currents of thought in Latin

As Karen O'Brien, a scholar of environmental change, argues, climate adaptation ultimately requires a shift in imagination – the ability to recognise that social and political systems can change as profoundly as ecological ones. Climate justice scholar Farhana Sultana similarly emphasises that adaptation grounded in justice is not only about surviving worsening extremes, but building the relational, emotional, and political capacities required to live well together as conditions deteriorate. None of this is straightforward. Intensifying weather extremes, accelerating displacement, and the re-drawing of ecological boundaries will test societies in unprecedented ways. There are no guarantees. Regeneration cannot always undo ecological damage, nor can adaptation prevent all forms of dislocation.[67]

Holding on to what is possible – the continuation of life in place – will require deep structural change. Advocates of transformational adaptation argue that this will involve changing the underlying systems that produce vulnerability, rather than reinforcing them. Geographer Mark Pelling and environmental change scholar Karen O'Brien both frame this as the point at which adaptation becomes an explicitly moral and political act – a shift in how societies understand risk, responsibility, and possibility. Prof Siri Eriksen from the Norwegian University of Life Sciences and development scholar Jonathan Ensor argue that adaptation becomes transformative when it is anchored in justice, reciprocity, and care: when it redistributes power, challenges inequality, and centres those most affected, rather than simply managing risk.[68] In this context, political action is not separate from care, preparedness or responsibility. It is one of the ways care is exercised at scale – by shaping the conditions under which harm is prevented, resources are distributed, and futures remain possible. As Bec insisted during our conversation, '*everyone has a role to play, and valuing the different skills is really*

America', *Agroecology and sustainable food systems*, 41/3–4 (2017), 231–37, <https://doi.org/10.1080/21683565.2017.1287147>, accessed 4 October 2025.

67 Karen O'Brien, 'Is the 1.5°C target possible? exploring the three spheres of transformation', *Current opinion in environmental sustainability*, 31 (2018), 153–60, <https://doi.org/10.1016/j.cosust.2018.04.010>, accessed 4 October 2025; Farhana Sultana, 'The unbearable heaviness of climate coloniality', *Political geography*, 99 (2022), article 102638, <https://doi.org/10.1016/j.polgeo.2022.102638>, accessed 4 October 2025.

68 IPCC, *Climate change 2022;* Mark Pelling, *Adaptation to climate change: from resilience to transformation* (Routledge, 2011); Eriksen, Nightingale and Eakin, 'Reframing adaptation'; Jonathan Ensor and Rachel Berger, *Understanding climate change adaptation: lessons from community-based approaches* (Practical Action Publishing, 2009), <https://practicalaction-publishing.com/book/2243/understanding-climate-change-adaptation>, accessed 6 October 2025.

important. It's only when we understand each other's roles that we can meet in the middle and make the connections'.

Pathways of resilience

We return here to the idea of resilience – a central yet contested idea in climate disruption discourse. Indeed, as Bec inquired earlier in this chapter: *'People can't assume that when you call Triple Zero, someone will come. So, what do we do instead?'*

It's a question full of big implications, not least about the role played by local people when the cavalry is not coming. This is where the concept of resilience comes in. It's a well-worn idea that has deep roots in the physical sciences. In the field of engineering, resilience came to refer to the capacity of a material to absorb stress and return to its original state – the classic idea of 'bounce-back'. In ecology, Canadian ecologist C. S. (Buzz) Holling used the term to describe an ecosystem's capacity to absorb disturbance without breaking down or collapsing altogether. As the term migrated into the social sciences and disaster studies, it came to describe how individuals and communities withstand and recover from crises. Disaster scholar Siambabala Bernard Manyena argues that by the early 2000s resilience was widely used to denote social recovery, but was also increasingly criticised for its vagueness and potential to obscure structural causes of vulnerability. Planning theorist Siobhán Davoudi similarly warns against the notion of 'bouncing back', noting that returning to pre-disaster conditions is neither realistic nor desirable when those conditions contribute to risk in the first place.[69]

Development scholars Christophe Béné et al. have proposed a framework that distinguishes between three interrelated capacities of resilience: absorptive, adaptive, and transformative. Absorptive capacity refers to the ability of people, organisations or systems to absorb shocks and maintain essential functions during and immediately after a critical event. Examples of this sort of responsiveness include emergency services, financial buffers, and community safety nets. Adaptive capacity involves learning from experience and making incremental changes to reduce future risk. This includes diversifying livelihoods, strengthening public

69 Crawford S. Holling, 'Resilience and stability of ecological systems', *Annual review of ecology and systematics*, 4 (1973), 1–23, <https://doi.org/10.1146/annurev.es.04.110173.000245>, accessed 4 October 2025; Siambabala Manyena, 'The concept of resilience revisited', *Disasters*, 30/4 (2006), 433–50, <https://doi.org/10.1111/j.0361-3666.2006.00331.x>, accessed 4 October 2025; Simin Davoudi, Keith Shaw, L. Jamila Haider, Andrea E. Quinlan, Garry D. Peterson, Cathy Wilkinson et al., 'Resilience: a bridging concept or a dead end. the politics of resilience for planning: a cautionary note', *Planning theory & practice*, 13/2 (2012), 299–333, https://doi.org/10.1080/14649357.2012.677124>, accessed 4 October 2025.

health systems, or redesigning urban spaces for heat resilience. Transformative capacity is the ability to fundamentally alter system structures, power relations, and cultural narratives to create new, just, and sustainable socio-economic and political arrangements. Unlike the notion of bounce back, adaptive resilience sometimes requires letting go of what existed before.[70]

One of the most important points to come out of the adaptation literature, however, is that civic responsiveness is critical in most disaster scenarios. This sounds like an obvious point for anyone who has witnessed such an event, but governments still struggle to understand it. Sociologists Daniel Aldrich and Mark Meyer show that policy continues to privilege physical infrastructure over social networks, even though decades of disaster research demonstrate that neighbours and informal groups are almost always the first responders after catastrophe. Classic disaster studies by Russell Dynes and Enrico Quarantelli also report that community-based, spontaneous organising consistently outperforms formal systems in the immediate aftermath of disaster. According to a number of resilience researchers,[71] local connections have been shown to enhance access to resources, emotional support, information and problem-solving. Key to this, as Fran Norris et al. point out, is a commitment to shared resources, supportive relationships, and collective action. A systematic review by Sheetal Patel and co-authors identifies nine common characteristics – vital in times of disaster – that typify resilient communities. These include shared knowledge, strong networks, effective communication, good governance and leadership, health, access to resources, economic investment, disaster preparedness, and mental outlook.[72]

---

70 Christophe Béné, Robert Wood, Andrew Newsham and Mark Davies, *Resilience: new utopia or new tyranny?* IDS Working Paper 405 (Brighton: Institute of Development Studies, 2012); Helen Jeans, Gonzalo E. Castillo and Sarah Thomas, *Absorb, adapt, transform: resilience capacities* (Oxfam, 2017); IPCC, *Climate change 2022.*

71 Krzysztof Kaniasty and Fran H. Norris, 'A test of the social support deterioration model in the context of natural disaster', *Journal of personality and social psychology*, 64/3 (1993), 395–408, <https://doi.org/10.1037/0022-3514.64.3.395>, accessed 4 October 2025; John J. Beggs, Valerie A. Haines and Jeanne S. Hurlbert, 'The effects of personal network and local community contexts on the receipt of formal aid during disaster recovery', *International journal of mass emergencies & disasters*, 14/1 (1996), 57–78, <https://doi.org/10.1177/028072709601400104>, accessed 4 October 2025.

72 Aldrich and Meyer, 'Social capital and community resilience'; Kaniasty and Norris, 'A test of the social support deterioration model'; Enrico L. Quarantelli, 'Emergent citizen groups in disaster preparedness and recovery activities' (1984), <http://udspace.udel.edu/handle/19716/1206>, accessed 4 October 2025; Jeanne S. Hurlbert, Valerie A. Haines and John J. Beggs, 'Core networks and tie activation: what kinds of routine networks allocate resources in nonroutine situations?', *American sociological review*, 65/4 (2000), 598–618, <https://doi.org/10.2307/2657385>, accessed 4 October 2025; Norris et al., 'Community

The 2022 Northern Rivers flooding events in New South Wales demonstrated these resilience characteristics in action – both before and after the disasters. Local residents, neighbourhood centres, volunteer groups, and mutual aid initiatives had already built strong relationships, shared skills, and local support systems that shaped how the region responded to challenges. People were thus able to organise quickly during the floods, and their contributions remained visible in efforts that followed. Community groups continue to support one another through recovery, adaptation, and future planning. As the authors of this book have witnessed, the Northern Rivers experience illustrates vividly how resilience arises when relationships, shared purpose, and local knowledge are already present, thereby enabling people to adapt and recover even in the face of institutional shortcomings.

But we need to be cautious here. As previously noted, an overemphasis on community resilience can lead to governments and institutions outsourcing responsibilities and costs which, according to scholar of neoliberal governmentality, Jonathan Joseph, reflects a neoliberal impulse to rely on citizens to 'self-organize' and 'cope' with disasters while the state steps back. Far from being a neutral term, and stripped of context and political meaning, the idea of resilience can easily become a tool for the powerful to devolve responsibility onto citizens. As Muriel Côté and Andrea J. Nightingale, scholars of socio-ecological systems point out, the idea of resilience is often used to portray crises as sudden, spectacular and unforeseen external shocks rather than the consequences of long-standing inequalities and harmful industrial and economic systems. It is a familiar line that serves the interests of capital while, in effect, seeking to pull the wool over peoples' eyes.[73] Community resilience is strengthened not only through preparedness and mutual aid, but through collective action that makes demands on institutions. When communities organise, campaign, and advocate together, they increase

resilience as a metaphor'; Shahrzad S. Patel, Michael B. Rogers, Richard Amlôt and G. James Rubin, 'What do we mean by 'community resilience'? a systematic literature review', *PLOS Currents Disasters* (2017), <https://doi.org/10.1371/currents.dis.31b5bccd32108faab5ee0f-f7f4f5799e>, accessed 4 October 2025; Russell R. Dynes, *Organized behavior in disaster*, The Disaster Research Center Series (Columbus, Ohio: Disaster Research Center, The Ohio State University, 1974).

73 Jonathan Joseph, 'Resilience as embedded neoliberalism: a governmentality approach', *Resilience* 1/1 (2013), 38–52, available at <https://doi.org/10.1080/21693293.2013.765741> (accessed 18 March 2025); Muriel Cote and Andrea J. Nightingale, 'Resilience thinking meets social theory: situating social change in socio-ecological systems (SES) research', *Progress in human geography* 36/4 (2011), 475–89, available at <https://doi.org/10.1177/0309132511425708> (accessed 18 March 2025).

their capacity not just to respond to disruption, but to influence the conditions that shape their exposure to harm in the first place.

Case study: The Community Carers and Responders (CCR) network

In recent years, official reviews of major disasters in Australia have underscored the limits of state-led emergency response and the critical role of communities in bridging those gaps. The 2020 Royal Commission into National Natural Disaster Arrangements concluded that, 'in significant emergencies and disasters, emergency management personnel do not, and never will, have the capability and capacity to solve the emergency threat for every individual at risk'.[74] In other words, there cannot be a fire truck in front of every house and official responders will not be on people's doorsteps in the immediate aftermath of a disaster. Understandably then, the 2022 New South Wales Flood Inquiry recommended the establishment of 'a community-based first responder training program' to ensure that residents are equipped to act when official resources are delayed or overwhelmed.[75] This policy prescription was already being acted out in the Northern Rivers region through Plan C's Community Carers and Responders (CCR) network from 2021, which offered a practical model for how local communities can organise, train, and mobilise collectively to strengthen preparedness, mutual aid, and recovery capacity – from the ground up.

According to Plan C, a not-for-profit registered charity founded in 2019 in anticipation of large-scale compounding disasters, the CCR network supports crisis prevention, preparedness, response and recovery at the community level by setting up 'a trained and connected network of leaders dedicated to community cohesion and resilience'.[76] At the heart of the network is a simple realisation: namely, 'I can't be safe if my neighbour is not safe, and they can't be safe if their neighbour is not safe'. This recognition forms the network's relational ethos – the understanding that resilience is not an individual achievement, but something built and sustained through interdependence. The CCR network embodies this principle by cultivating trust, reciprocity, and shared responsibility across

---

74 Royal Commission into National Natural Disaster Arrangements, 387

75 Mick Owens and Maryanne Fuller, *2022 NSW flood inquiry: final report* (Sydney: New South Wales Government, 2022), 20.

76 Plan C, *Community carers and responders* webpage (no author, no date), <https://www.planc.org.au/community-carers-and-responders>, accessed 4 October 2025.

communities. It reframes preparedness from a personal duty into a collective practice of care, where wellbeing and safety are mutually reinforcing.[77]

However, relationships alone are not enough. The CCR network also focuses on building skills so that any future disaster response is grounded in existing local capacity and trust – we need people with the knowledge and wherewithal to respond to practical challenges. Volunteers complete a multi-day training that covers disaster preparedness, community organising, emergency communications, food, water and energy security, first aid, and psychological first aid.[78] Participants then go on to lead local resilience projects with optional support and mentoring. The programme blends both practical and emotional learnings through the development of technical abilities such as risk assessment and communication alongside self-care, as well as empathy and collective problem-solving. By fostering peer support and emotional awareness, the CCR network recognises that resilience depends as much on emotional intelligence and situational awareness as on logistics and plans. Participants often report that they have benefitted not just from skills development, but from sense of belonging and confidence that comes from knowing they are part of a caring network rather than facing future calamities alone.[79] Since its launch, the CCR network has trained several hundred community members and produced numerous tangible outcomes across NSW and Queensland. These include the formation of neighbourhood resilience groups, the development of risk-reduction plans, the establishment of radio/emergency communications networks, food-security initiatives, and many community-building and educational events.[80]

While the CCR network demonstrates the possibility of community-led resilience-in-action, it also lays bare some of the limitations and tensions of such an approach. First, community-led resilience does not replace formal state responsibilities; indeed, as previously noted, there is always the risk of government withdrawal or an operating assumption that community actors will fill the gaps. Second, participation in civic action is often patchy and uneven: not all communities have equal resources, networks or capacity to engage in disaster

---

[77] Plan C, *Community carers and responders* webpage.

[78] Lara Bell, 'Plan C – strengthening communities through free resilience education' (20 August 2023), *The Lismore App*, <https://lismoreapp.com.au/NewsStory/plan-c-strengthening-communities-through-free-resilience-education/64e042d2f5c8a30028e1612b>, accessed 4 October 2025.

[79] Plan C, *Community carers and responders* webpage.

[80] Plan C, *Community carers and responders* webpage.

preparation – and not everyone in any given locality will agree on the scale of the threats before us (if any, according to some) and how to deal with them.

Nonetheless, the CCR network illustrates how the absorptive–adaptive–transformative pathways of resilience can be operationalised. Training and connection-building corresponds with absorptive capacity (preparing for shocks, building local cohesion), while neighbourhood projects, risk-reduction plans and resource networks reflect adaptive capacity (adjusting systems, preparing for changing conditions). Finally, the deeper orientation towards local leadership, relational care, strengthening food, energy, social systems – and the idea of regeneration rather than mere recovery – all gesture toward transformative capacity.

Through mutual aid, shared learning, and an ethos of interdependence, CCR participants are not only preparing for future disasters but actively redesigning the social architecture that sustains them. In this sense, CCR is a prototype for the regenerative communities to come: decentralised, compassionate, and alive to the reality that the pathway to safety is civic solidarity.

Regenerate everything

After years of compounding crises – drought, bushfires, floods, pandemics, economic disruption, housing shortages, and widespread mental distress – and given what the science is telling us, there can be no more 'business as usual' in the post-Anthropocene era. While 'bouncing forward' has become a popular phrase in resilience policy, genuine transformation implies confronting and changing the root causes of vulnerability and risk – for example the growth imperative of capitalism, extractivist logics, the reification of nature, overshoot and the colonial legacies of dispossession.

Regeneration in this sense is literally revolutionary insofar as it upends all the exploitative and reified practices associated with extractivist capitalism. Its call for the redesign of economy, food, water, energy, waste, housing, transport, health, education, and even our shared values – are aimed at sustaining and replenishing life rather than depleting it. Accordingly, regeneration means designing systems that leave more than they take, creating abundance through cooperation and connection, enabling us to live within ecological and social limits. If natures gifts are to be reciprocated then we have little choice but to Care for Country. In seeking to restore vitality and enhance the capacity of people and nature to co-evolve in mutually beneficial ways, we need to think and act in these reciprocal ways. Regenerative design practitioner, Pamela Mang, and Bill Reed, regenerative development theorist, describe this orientation as 'designing with living systems'. This moral investment in truly reciprocal ways of eyeing have huge payoffs for all ecologies. As geographer Samuel Buckton and colleagues point out, regenerative social-ecological systems cultivate diversity, reciprocity and resilience – they literally give and give back in ways that enable life to flourish. Simply taking has no place in this context.[81] Because regeneration challenges extractivist value systems, it inevitably enters political terrain, as even while regenerative practices often emerge locally, they cannot scale-up or endure without institutional support. Energy systems, education, finance, land use, and labour markets are governed at national and transnational levels. Without political engagement to reshape these structures, regenerative initiatives risk remaining marginal – inspiring, but structurally constrained.

Similarly, a regenerative economy challenges the assumption that economic prosperity must depend on perpetual extractivist growth, and instead asks how wealth, energy, and resources can circulate equitably within planetary boundaries. Economist Kate Raworth, known for developing the idea of 'doughnut economics' at Oxford University's Environmental Change Institute, argues that the goal of economic design should be to meet everyone's needs while staying within the planet's life-supporting boundaries. The inner ring of the doughnut thus represents life's essentials – food, housing, healthcare, education, equity, political voice, while the outer ring represents the planetary boundaries of climate stability, biodiversity, fertile soils, clean water, and the ozone layer. Overshooting these boundaries, says Raworth means environmental breakdown – a point well established in Chapter 2.

Likewise, Tim Jackson, a sustainability economist at the University of Surrey, proposes an economy built around 'prosperity without growth' – one that values health, equity, and purpose more than output. Other economists, such as Herman Daly, a pioneer of ecological economics, have long called for a 'steady-state' economy that maintains resource use at levels compatible with the Earth's capacity to regenerate. As biologist and Natural Intelligence expert Leen Goriseen explains: 'it's not that complicated: we need to replace the self-destructive logic of our Western economy with the life-enhancing logic of living systems. And that begins with our worldview'. She adds: 'in nature, every part adds value to the whole it belongs to – and in doing so, it receives nourishment in return. That is the foundation of value creation that can endure over the long term'.[82]

81 Pam Mang and Bill Reed, *Regenerative development & design: a framework for evolving sustainability* (New Jersey: Wiley, 2016); Sarah Buckton, Ioan Fazey, Beth Sharpe et al., 'The regenerative lens: a conceptual framework for regenerative social-ecological systems', *One Earth*, 6/7 (2023), 824–42.

82 Leen Gorissen, interview with wearethe_regeneration, Instagram post (18 January 2026).

These frameworks suggest that investment, policy, and innovation should focus on restoring the ecological systems that underpin life – forests, soils, water, and communities – rather than ceaselessly extracting from them. Examples of such thinking are emerging in practice. Amsterdam has adopted Raworth's doughnut model to guide its urban planning, aligning social inclusion with ecological boundaries. Costa Rica has embedded forest restoration and renewable energy into its national development strategy.[83]

In agriculture and land management, regeneration is evidenced in practices such as composting, planned grazing, diversified cropping and soil restoration – approaches that build soil carbon, improve water cycles, support biodiversity and strengthen livelihoods. For instance, agricultural geographer Hannah Gosnell, environmental scientist Nicholas Gill and sustainability scholar Michelle Voyer document how such transitions occur on farms shifting toward regenerative systems.[84]

A parallel application of regenerative principles can be found in area of the built environment. Ecologist and urban futurist Herbert Girardet, for instance, argues that 'regenerative cities' must move beyond sustainability to become systems that give back more than they consume. Advocates of regenerative development, Pamela Mang and Bill Reed, likewise reconceptualise built environments as living systems embedded in their bioregions, arguing that settlements should be viewed as being embedded within larger ecological systems. The health of a city, they contend, depends on the vitality of its bioregion – its watersheds, soils, biodiversity, and cultural traditions. This form of 'living system thinking' encourages the joint evolution of settlements and natural systems, so that population centres emerge from the specific qualities of the bioregion rather than from preconceived generic 'green' strategies.

These approaches have another major practical benefit, which is to shift thinking away from retrofitting damaged structures toward designing neighbourhoods, buildings and infrastructure that support both ecological and human regenerative development.[85]

---

83 Raworth, *Doughnut Economics*; Tim Jackson, *Post growth: life after capitalism* (Cambridge: Polity Press, 2021); Herman E. Daly, *Ecological economics and sustainable development: selected essays of Herman Daly* (Cheltenham: Edward Elgar Publishing, 2014).

84 Hannah Gosnell, Nicholas Gill and Michelle Voyer, 'Transformational adaptation on the farm: processes of change and persistence in transitions to 'climate-smart' regenerative agriculture', *Global environmental change*, 59 (2019), article 101965, <https://doi.org/10.1016/j.gloenvcha.2019.101965>, accessed 4 October 2025.

85 Herbert Girardet, *Creating regenerative cities* (Oxon: Routledge, 2014); Mang and Reed, 'Designing from place'.

At the level of governance, regeneration requires decision-making grounded in justice, participation and long-term care. Transformative governance models emphasise local knowledge, multi-stakeholder collaboration and institutional accountability – approaches that ensure the adoption of the principle of custodianship, rather than simply that of management. The goal is not just to live within limits but to live in right relationship with the living world that makes flourishing ecological systems possible. Agency takes the form of refusal: refusing narratives that normalise extraction, refusing policies that sacrifice some for others, and refusing to accept that harm is inevitable when it is actively produced.

Regeneration, then, is not an abstract aspiration or a future state to be attained, but the conduit through which action becomes coherent. It is the context that gives meaning to everyday practices – from community care to political organising, from skill-sharing to economic redesign – and that allows responses to collapse to be oriented toward justice rather than mere survival.

From survival to renewal

At this point in our argument, it almost goes without saying that genuine regeneration cannot occur while extraction continues unchecked. Without decisive emissions reduction, efforts at regeneration risk becoming gestures of repair on an otherwise collapsing system – the shifting of chairs on the Titanic was always going to be futile. Conversely, mitigation without regeneration risks becoming a project of naked austerity – reducing harm without reimagining how we should live – and we *do* need to live very differently in this time of endings. The two – mitigation and regeneration are inseparable: mitigation halts the bleeding; regeneration restores the pulse.

Further, while mitigation addresses one of the main causes of our planetary predicament; adaptation faces the consequences; resilience sustains us through disruption; and regeneration rebuilds the foundations from which a more-than-bearable life becomes possible. This might involve decarbonising transport, restoring wetlands to buffer floods, building social networks that share resources during crises, and rewilding degraded landscapes to restore biodiversity. This shift from crisis management to systems change, from defensive adaptation to creative redesign is indispensable for a liveable future.

Against all odds

As Bec shared at the beginning of this chapter, '*I arrive [to the community meeting] exhausted, and I come out buzzing... this is the work that needs to happen*'. Readers are encouraged to find the intersections between their abilities and their concerns – we have more to say on this in the Conclusion, but as Bec asserts, '*we have to do something about this*'. For those who are committed to regeneration and systemic change, they must operate within a contested space where change is often met with denial, derision, or hostility. Change agendas can be dismissed as unrealistic, elitist, or threatening to established interests, while apathy can erode momentum and undermine collective will. Sustaining personal and civic engagement therefore requires emotional fortitude, moral clarity, and empathy: recognising that opposition and indifference often stem from fear, fatigue, or loss of trust. Effective change agents must learn to listen as much as they advocate, finding entry points that connect shared values – security, dignity, and continuity of life – to broader ecological and social transitions. In so doing, they can promote and model the relational and regenerative ethics that are vital to very futures they are seeking to build. In practice, this means forging alliances across difference – working with farmers to restore soils, with workers to create secure 'green' jobs, with communities to design local energy and food systems, and with policymakers to embed long-term, systems thinking into governance. By working shoulder to shoulder with others – not always easy, we know, but vital in turbulent times – it is possible to see regeneration as a shared pathway to stability, dignity, and collective survival. New, collaborative ways of thinking are needed, just as we need to learn practical skills, cultivate calm, and build mutual trust so that when disasters strike, people and communities will be prepared. It is through social and political movements that individual concern is translated into collective pressure. Movements provide the infrastructure through which people coordinate and contest power over time – whether around climate justice, workers' transitions, biodiversity protection, housing security, or disability-inclusive adaptation. Without such vehicles, agency remains fragmented and easily absorbed.

Readers might want to reflect on the influence of their everyday choices – how they vote, spend, work, and speak – and how the choices they make align, or not, with a vision of justice, compassion, and regeneration. Challenging extractivist systems requires participation beyond private life – in civic forums, institutions, movements and struggles where the terms of our collective future are negotiated. What this participation can look like, at different scales and capacities, is taken up in the Conclusion.

Regeneration also demands a cultural shift that is often left unspoken: restraint. Not restraint as austerity or moral purity, but as the deliberate choice to live within limits – to distinguish between what is desired and what is needed. In societies shaped by extractivism and over-consumption, excess has been normalised and marketed as freedom. Regenerative futures, by contrast, require relearning sufficiency. This raises uncomfortable but unavoidable questions about consumption, mobility, and reproduction – about how much is enough, and who decides. Such questions cannot be answered through individual willpower alone. They require political frameworks that make restraint possible without harm: affordable public transport, accessible local food systems, secure housing, and social supports that allow people to choose less without bearing disproportionate costs.

Finally, you might ask what practices genuinely restore *your* sense of balance, belonging, and vitality in a changing world. These could be moments of quiet reflection, time in nature, shared meals, or acts of generosity that reaffirm connection and care. Or it can be removing unnecessary burdens from one's life, shedding the busyness of daily life. You might therefore ask how can I design a life that is more resilient and regenerative? How can I contribute to the wellbeing of my community and ecosystem, supporting not only human life but the wider web of beings that sustain it? What steps can I take to reduce my ecological footprint – not from guilt, but from love for the places and relationships that give life meaning?

Conclusion

Responding effectively to the Great Unravelling requires an integrated framework that holds mitigation, adaptation, resilience, and regeneration as mutually reinforcing. Mitigation remains an ethical imperative: to reduce future harm and preserve the possibility of life for coming generations, while adaptation demands humility and inclusiveness, recognising that effective responses must engage with cultural and political realities, not merely technical solutions. Resilience on the other hand, underscores the necessity of strengthening our individual and collective capacity to survive and evolve through disruption, while regeneration gestures towards reimagining the social and ecological systems that sustain and nurture all life.

Living well in this time of disruption is not about returning to certainty but to nurture belonging to a future we are still capable of shaping. The Unravelling, then, is not merely an ending but an invitation to live very differently – we really do have little choice. In essence, what we have called for is a renewed re-engagement in civic life in the context of disruption and upheaval. We have stressed the importance of social solidarities and community resilience, not as feel-good shibboleths, but as core in a united effort to live as well as we can in an upended world. Individuals, households, neighbourhoods, communities, institutions, states and the Commonwealth are all integral to this project of renewal which should be grounded in justice, Indigenous sovereignty, and solidarity with those already on the frontlines of disruption.

If hypernormalisation thrives on silence, fragmentation, and the privatisation of responsibility, then agency must be collective, vocal, and structural. The practices explored in this chapter – mitigation, adaptation, resilience, and regeneration – are not ends in themselves, but starting points for action. The question that remains is how these ideas translate into daily life, civic engagement, and political pressure. That is where we now turn.

Our intention is to get first-person personal, so: based on what I have read and what I already know and feel what can I now do? Once the pause is over, or once I see that modernism has been hospiced, how do I then proceed? What steps can I take to make some sort of meaningful difference? Let's reflect on this, together.

Conclusion: Facing endings, cultivating beginnings

The eyes of the future are looking back at us, praying that we will see beyond our own time.

—TERRY TEMPEST WILLIAMS

Another world is not only possible, she is on her way. On a quiet day, I can hear her breathing.

—ARUNDHATI ROY

You don't need to do everything. Do what calls your heart; effective action comes from love. It is unstoppable, and it is enough.

—JOANNA MACY

Retracing the journey

For a book titled *Unspeakable* we have certainly expended a lot of words. We began with a simple yet deeply unsettling assertion; namely, that we are living in a time of endings. The ecological stability that once allowed human societies to flourish is unravelling before our eyes, and with it the political, economic, and cultural certainties that once anchored everyday life. In Part I, we traced how this disturbing situation occurred – not as a sudden rupture, but as the cumulation of centuries of extractivism, colonial expansion, growth economics, and the hegemonic silence that allowed this system to endure. Denial, distraction, compulsory optimism, and the pervasive belief that someone, somewhere will eventually fix things. All these smoke and mirrors have blunted our understanding of the dangers we now face.

Through personal stories, scientific studies, and social critique, we have argued that the deteriorating biosphere – one of the key drivers of planetary destruction – is not only an imminent threat, but a crisis of humanity. It unsettles our sense of place, of who and what we are, and our purpose and direction in life. At its core, the climate catastrophe is an existential rupture that beckons us to reckon with grief, fear, and uncertainty.

The unspeakability of our predicament – the manufactured silence that makes it difficult to name the climate disaster for what it is – has occupied our attention in the early part of the book. We considered how this silence is maintained by avoidance and fear, as well as by the stories that modern societies tell themselves about progress, growth, and stability. It is also a silence prompted by the emotional overwhelm that comes with the dark knowledge of climate disruption; and of course, by systems of power that depends on obscuring the reality of our situation. As we argued in Chapter 3, hypernormalisation creates the illusion of continuity – the steady march of the everyday – even as evidence of collapsing ecological systems mount. Citizens are encouraged to perform normality rather than acknowledge what is happening around them. This has real world consequences. In effect, hypernormalisation is a direct ideological hindrance to action, influencing what can be imagined, or politically demanded – it keeps us stuck in a dangerous illusion. Naming the climate catastrophe, and the end of industrial modernity, is therefore key to reclaiming agency – a first step toward breaking the spell of inevitability and widening the space for truth, grief, and transformation.

In Part II, we turned inward and outward – exploring how individuals, households, and communities can live meaningfully in a time of upheaval. We focused on the emotional labour required of this moment: grief that signals attachment to what is being lost; anger that, when named and channelled, can sharpen moral clarity; despair that, when honoured, can become the foundation for agency – rather than paralysis. Through the prism of radical acceptance – a willingness to remain present to reality and to act from that place – we have shown that acceptance enables us to act from steadiness rather than panic, while deep peace is framed, not as escape or resignation, but as an inner state that fosters stillness amid turbulence.

From there, the book discussed the meaning of community resilience within the context of a reinvigorated civic society. We showed how relationality, built through networks of trust, reciprocity, and shared purpose rather than rugged individualism, form the backbone of resilience. Communities built on this shared quality become lifelines of survival and renewal in a destabilised world. Through personal stories, and examples from across Australia and beyond, we illustrated how people already hold the line when institutions falter in times of disaster. Mutual aid and embodied care emerged as vital expressions of agency – precursors to the bigger changes needed to navigate our way through turbulent times.

In looking at the more practical responses to climate disruption, we argued that mitigation remains a moral imperative, not because it will restore the world we knew, but because every fraction of warming avoided will reduce suffering. We showed too why adaptation – an unavoidable necessity – demands more than infrastructure and engineering. Above all, it necessitates cultural change, justice, redistribution, and long-term thinking in a political environment that is often averse to such things. Our discussion drew attention to various adaptation measures including relocation, food and water security, new infrastructure, and the redesign of transport, production and other systems. Mitigation and adaptation measures, we argued, should sit alongside regenerative practices that allow life to flourish – one without the others is meaningless.

Across these chapters, we have emphasised that the challenge of our time is not merely technical but moral, emotional, relational, and political. The challenge is not only how will we survive? but *how shall we live?* What does it mean to care for one another when the old order fails? What becomes possible when we accept the gravity of our situation without surrendering to hopelessness? How do we rediscover the civic imagination and solidarity needed to respond with dignity and care rather than fear?

This conclusion does not offer a programme to be followed or a checklist to be completed. It gathers together the threads of the book and points toward forms of action and engagement that have emerged in response to the conditions described, leaving readers to consider how – and whether – these might take shape within their own lives and contexts. We have been careful to make no promises of rescue. Instead, we have proposed responses that might help us to chart a path through turbulent times. Ours is an invitation to honesty, to courage, and to renewed forms of agency grounded in care and empathy – for ourselves, each other, and Country. The 'so what?' of this book is therefore a call to inhabit this moment fully: to face what is ending, to nurture what remains, and to begin reimagining and shaping what might still be possible.

What you can do

Any book that lays claim to change, we believe, should offer some signposts as to what the reader can do to contribute to this process. We certainly have not got all the answers – not by a long stretch – but there may be some useful takeaways to consider.

As we turn toward the question of what can be done, it is worth recalling the personal stories that have accompanied this book, as reminders of the agency we all have. Cat's response to the Lismore floods, Gin's long custodianship of forest and fire, Bobbi's practice of attention and stillness, and the many acts of mutual aid that unfolded after disaster all show that meaningful action rarely begins with certainty or grand design. It begins with care – for people, for place, for what is being lost and what still matters. These stories remind us that agency is not reserved for the heroic or the prepared, but emerges wherever people choose presence over withdrawal and responsibility over resignation. The invitations that follow are offered in this same spirit, as ethical signposts, asking how you might respond, here and now, in ways that are faithful to your own values, capacities, and relationships.

None of what we suggest is easy or even doable, and you may, naturally, have your own ideas about how and where you would like to devote your energies. When it comes to big systems change, there are many obstacles to overcome, not least the intransigence and rootedness of hegemonic power. As discussed in Chapter 3, entrenched systems of power – driven by political short-termism and corporate greed – actively reinforce the status quo, so a call for systemic change can appear idealistic, naïve even.

Yet change remains both possible and necessary, and in many ways, it's already happening. Take a look around. Just consider the millions of people across the world who have adopted 'alternative' ways of relating, knowing and being – all reflecting values and practices set out in the last chapters. (We are certainly not the first to write about different ways of doing things). Consider too how the climate catastrophe has increased awareness of the system that has led us to this point. Yes, the dominant neoliberal system endures – for now – but as ecological, social, political and economic disruptions converge, transition will happen – either by design or by disaster, or both. The challenge then is to design our shared responses by inserting justice, resilience, and regeneration into policies, institutions, and daily life.

Waiting for permission to act is no longer viable, or wise; the work must begin wherever we are. What follows, then, is not a list of prescriptions – although this can be helpful – but an invitation to weave together insights that might console and give us direction. Suffice to say: to live well, or as best we can, in the face of disruption and chaos is to pay heed to our interior worlds, to build democratic and socially just solidarities, and to pursue our desire for a better world by participating in systems change.

Beginning with ourselves

As explored in Chapter 4, the capacity to act meaningfully in the midst of disruption begins in the inner life. In a world saturated with distraction and manufactured consent, tending to our own attention becomes part of the work: the quiet act of seeing clearly, resisting numbness and refusing to be swept away by doom scrolling, infotainment and the noise of the moment.

To face the Great Unravelling without denial or resignation, we must cultivate the ability to remain present with what is – uncertainty, anguish, and grief, but also with beauty, solidarity, and the possibility of renewal. Radical acceptance then, enables us to see our predicament clearly, without falling into despair, while grief work helps us honour what has been lost while remaining open to what is still possible. When tended carefully, grief becomes a connective force rather than a private burden – a reminder of why we care at all.

As you, the reader, consider this, you might entertain the thought of participating in, or deepening, practices that cultivate presence – journalling, contemplative breathwork, attentive walking, crafting, making art or music, observing seasonal shifts, listening to Country, or carving out moments of silence and reflection. These will help steady your attention and support emotional regulation. The benefits accruing from these practices accumulate over time, generating the inner spaciousness from which clarity, compassion, and courage emerge. The effort is worth it. From this too, you may come to know the wonders of deep peace, a cultivated calm that allows discernment and grounded response instead of frantic reaction: stillness in action.

You might also want to think more deeply about the relational philosophy of inter-being, associated with Thích Nhất Hạnh and resonant with many Indigenous worldviews. Inter-being reframes agency by reminding you that we all exist in relationship: with ourselves, others, with Country, with the more-than-human world. You may find that this will loosen the grip of individualistic hyper-responsibility – the sense that one must single-handedly 'fix' the climate predicament – while strengthening your obligation to act with and for others. The sense of agency you will come to experience, therefore, is always situated and relational.

Alongside collective political action, many people find themselves confronting practical questions about everyday life. These are not questions of individual virtue or moral purity, but of alignment – how to bring daily practices into closer relationship with the realities and values explored throughout this book. For those with the capacity to choose, this can involve difficult conversations about restraint: about food systems, patterns of consumption, mobility, and reproduction. Questions about driving and flying, about the accumulation of goods, about how much space, energy or convenience is enough, or about the number of children one hopes to raise, touch deeply held desires and identities. They also sit uneasily within cultures that equate freedom with limitless choice. Engaging these questions honestly does not mean treating all choices as equal or ignoring the vast inequalities that shape people's options. Responsibility is not evenly distributed. Those who consume the most, travel the most, and benefit most from extractive systems carry perhaps a greater obligation to reduce harm. For others, especially those already constrained by poverty, disability, colonial dispossession or displacement, the problem is not excess but injustice. Nor can restraint be sustained through individual effort alone. Choosing less is only possible where social and political conditions support it – where public transport is viable, nutritious food is accessible, housing is secure, care is shared, and futures are not made precarious by policy design. Without these supports, calls for restraint risk becoming another form of quiet sacrifice imposed on those with the least power. Held in this way, restraint becomes neither punishment nor retreat but a form of care – for others, for future generations, and for the living systems on which all life depends. It is one way of withdrawing consent from extractivist logics that promise satisfaction through accumulation, and of making space for lives organised around sufficiency, relationship, and shared responsibility. For many, this work begins in conversation: with ourselves and with others, reflecting on what truly matters within the context of our own lives, responsibilities, and relationships.

Crucially, cultivating your inner resilience also means acknowledging one's own limits and embracing rest. In a time swamped by urgency, it is easy to imagine that meaningful care demands constant action. Yet sustainable engagement depends on pacing – you can't do everything, and nor should you try. Rest is not withdrawal but a form of refusal against extractivist logics that measure your worth through productivity and output. It is not an escape from responsibility but the practice that allows you to continue in the long-term. It might not be possible to be idle when you have a mortgage to pay, but as Bertrand Russell observed long ago,[1] slowing down, contemplating nature and life's wonders, are acts of resistance to the dominant order.

---

1 Vivian Marie Lewis, 'Learning from Bertrand Russell in today's tumultuous world' (14 May 2018), *The Conversation*, <https://theconversation.com/learning-from-bertrand-russell-in-todays-tumultuous-world-95405>, accessed 4 October 2025.

Strengthening relational and community resilience

By now, it should go without saying that resilience emerges most resonantly in and through relationships. Disaster research consistently shows that communities with dense social ties, shared knowledge, and established trust networks recover more quickly and are better able to withstand subsequent shocks. While at first it may feel intimidating, and perhaps even draining, to reach out to someone you don't know, it is more often than not, worth it. There are other things you can do. For example, you can begin to strengthen social ties by attaching yourself to neighbourhood communication networks; participating in mutual aid groups, bulk-buying arrangements, community kitchens, or local food cooperatives; or by contributing to community gardens, seed libraries, compost hubs, and Landcare groups. Or just spending time with others – hanging out, as teenagers do – without screens to distract us from genuine connection, is gold. The ensuing attachments will likely have demonstrably positive effects on you and improve your overall sense of wellbeing. By the same token, you might consider becoming involved in helping to establish resilience hubs that provide shared tools, emergency kits, information, and coordination. Initiatives such as the Community Carers and Responders (CCR) network, which you might look out for, demonstrate how skills and trust can be built well before disaster strikes, enabling you, your friends and acquaintances to mobilise quickly when institutions become overwhelmed.

These connective networks will likely spark your political imagination, offering glimpses of non-extractive ways of living that may guide future transitions. Even small acts of connection – checking on your neighbours, sharing skills, offering transport, providing informal care – will potentially give you satisfaction as well as strengthen your social attachments. These sorts of micro actions can often lead to more substantial forms of adaptation response. While your involvement is vital in building social capital and civic purpose, not all people are willing or able to step up in this way. It thus remains essential to acknowledge that relational capacity is unevenly distributed: time, health, disability, structural racism, poverty, and trauma shape who can participate, and how. Community resilience must therefore be supported by institutional investment rather than used as a justification for state withdrawal.

Civic agency also involves cultivating the emotional and relational conditions that allow collective action to flourish. Many people avoid talking about climate breakdown because the grief, fear or uncertainty feel too heavy. Normalising these conversations – in workplaces, friendship circles, schools and local groups – helps counter the isolation that erodes political courage. It also opens space for shared problem-solving and reduces the stigma of naming the truth. As discussed in earlier chapters, responsibilisation – the offloading of risk onto individuals and communities – is a common tactic in neoliberal adaptation policy. Local, community-based action cannot and indeed, should not replace obligations that are the preserve of the state. The role of the latter is to carry the heavy load by creating the conditions in which people and communities can feel relatively safe and supported and ultimately, prepared for when disasters strike.

Leaning into Indigenous relationality and custodianship

Indigenous knowledge systems offer important, life-giving insights into living well in a destabilised climate. Above all, First Nations cultures understand Country as a living, sentient, relational entity – a web of reciprocal obligations between people, land, waters, sky, and the more-than-human. As Linda Tuhiwai Smith notes, collective wellbeing begins when we 'remember who we are in relation to everything else'.[2] For non-Indigenous readers, perhaps like you, and us, engaging with Indigenous relational ontologies requires humility, reciprocity, and respect. It involves building relationships with Traditional Custodians; learning about the histories of dispossession and extraction that have shaped contemporary landscapes; supporting Indigenous-led land restoration, cultural burning, and governance initiatives; and advocating for the substantive inclusion of Indigenous ecological knowledge in adaptation, disaster risk reduction, and land-use planning.

However, learning from Indigenous knowledge must not become another form of extraction. Some practices cannot be adopted outside their cultural, spiritual, and legal contexts. If you are non-Indigenous but want to actively support sovereignty, land justice, and the transfer of decision-making authority, it is meaningful to not merely 'learn from' but to be in relationship with and part of. It is also important to acknowledge that engaging with Indigenous custodianship calls for a commitment to full spectrum change that recognises and decouple from ongoing colonial mentalities. This necessitates structural change, redistributive justice, and an ongoing acknowledgement of right relationship, not simply participation in discrete projects, even though it may start there.

---

2 Smith, *Decolonizing methodologies.*

Developing practical competencies and decentralised resilience

In an era of disruption, practical skills become vital aspects of resilience. Yet these must not be seen as triumphal acts of individual self-sufficiency but rather, as contributions to shared autonomy and mutual support. Which means, that if and when you choose to participate in skills-based learning, you'll be contributing to a shared, cooperative enterprise. Working alongside others, is just that, and in the context of disruption it's the best way of feeling supported and part of things.

Competencies such as growing food, maintaining soil health, saving seed, harvesting rainwater, reusing greywater, repairing tools, maintaining radios, mapping local risks, or coordinating neighbourhood logistics help decentralise resilience in ways that complement, rather than replace, institutional responsibilities. Practical competencies do not eliminate risk; they simply reduce vulnerability and expand the range of possible responses.

An important aspect of work that you might want to be part of is 'shelter resilience'. This is about enhancing the overall capacity of homes, workplaces, and shared refuges to withstand and protect against climate-related events. This may involve maintaining and keeping an eye on structural integrity of premises; improving insulation, ventilation, and passive habitability; strengthening critical systems such as power, water, waste, and communications; creating resilient landscapes; planning reciprocal shelter arrangements with neighbours or extended family; and ensuring redundancy across key household systems. Community-scale assessments of building vulnerabilities and advocacy for resilient standards, retrofitting programmes, and climate-responsive planning are essential parts of this work.

You might also consider involvement in other vitally important activities in areas to do with food, water, energy and communication. Or contribute to promoting health and wellbeing through community-supported agriculture, rainwater harvesting, microgrids, passive cooling and heating, low-energy devices, early-warning systems, community first-aid training, psychological first aid, community connectors, and neighbourhood evacuation planning. But these actions are unevenly available due to financial, geographic, and social constraints. It is therefore preferable to approach them without moralising or prescription.

Shaping systemic and structural transformation

If you've taken a deeper dive into the origins of environmental degradation you might wonder how to set about this big thing called structural transformation. While individuals cannot remake political economy alone, they can join the social and political movements that push for change. The most obvious way in a liberal democracy like Australia is to vote for candidates committed to fossil-fuel phase-out, support First Nations justice, ecological regeneration, and equitable adaptation. You might also participate in campaigns for halting new fossil-fuel projects, calling for the end of fossil-fuel subsidies or backing reforms that limit political donations and lobbying; support climate litigation; expand the use of renewable energies; advocate for ecosystem protection laws and rights-of-nature frameworks, fair insurance systems, and just relocation policies. You might also want to join with others in pushing for education reforms that integrate ecological literacy and adaptation across all curricula.

Equally important is the work people can do from where they already stand. Teachers can integrate climate justice into their curriculum; health workers can advocate for heatwave preparedness and highlight the health impacts of air pollution; engineers, architects and planners can press for resilient infrastructure; artists and storytellers can reshape public narratives; public servants can call out flawed processes or obstructive review pathways; and business owners can shift procurement away from suppliers that rely on exploitative or carbon-intensive practices. Such actions are modest in isolation but powerful in aggregation, especially within professions that shape public policy and community practice.

When enough people act, these interventions constrain the social licence that allows harmful industries to operate with impunity. For example, community pressure has already driven major councils to switch to renewable energy purchasing, super funds to tighten climate thresholds, and local banks to cut ties with high-risk fossil-fuel clients.

You can also contribute to alternative economic models that challenge neoliberal norms. Community wealth building – through cooperatives, community land trusts, employee-owned enterprises, and shared infrastructure – help redistribute value and reduce dependency on extractive markets. You might also want to take a close look at circular economy initiatives, time banking, non-monetary exchange, and local currencies, all of which strengthen local autonomy and resilience. You might further consider choosing banks, superannuation funds, and insurers that divest from fossil fuels and reinvest in renewables, social housing, and circular manufacturing shifts capital flows towards regenerative purposes. This can make a real difference. Advocating for wellbeing budgets, doughnut economics, and steady-state governance further supports transitions away from growth-dependent systems.

Yet it is critical to recognise that alternative economies cannot flourish without state involvement. Localism alone cannot overcome structural inequalities or decarbonise national energy systems. Collective political pressure is essential for institutional realignment – this is why it is important to consider carefully where you devote your energies. It's worth remembering too, that systemic change rarely occurs voluntarily; rather, it emerges through sustained and well-organised organised campaigns.

Climate and social justice movements provide vehicles for this collective pressure. Joining organisations focused on climate justice, workers' transitions, biodiversity protection, disability-inclusive adaptation, or renters' rights enables individuals to act on what they consider to be socially and politically just. Supporting global movements – including Pacific climate diplomacy, the Fossil Fuel Non-Proliferation Treaty, and loss-and-damage initiatives – aligns local action with transnational justice. The spirits of those fighting the enormous injustices of latter-day climate-colonialism are often raised through cross-border solidarities.

Contributing to ecological, social, and cultural regeneration

Participating in creek restoration, revegetation, habitat protection, regenerative agriculture, or urban greening contributes to ecological renewal. But regeneration demands more than ecological repair. It requires redesigning economic, social, and cultural systems so they replenish rather than deplete the conditions of life. Engaging in citizen science fosters literacy and accountability. Supporting cooperatives, local enterprises, repair and reuse initiatives, tool libraries, and resource recovery systems reduces material throughput and strengthens community autonomy. These are all good things to do in and of themselves.

Social and cultural regeneration are equally vital. This includes supporting mental health and wellbeing initiatives, and better access to primary health care; creating or participating in forums for dialogue, learning, and democratic engagement; mentoring young people and fostering intergenerational knowledge exchange; and nurturing artistic and cultural projects that reimagine futures beyond extraction. Cultural work – storytelling, art, ceremony, joy! – helps shift collective consciousness, enabling people to imagine life after the exhaustion of neoliberal modernity. As Indigenous scholar Tyson Yunkaporta observes, renewal begins in the stories we tell about who we are in relation to each other and to Country.[3]

Crucially, no one is expected to undertake all of these actions, nor are they universally accessible. Agency is always constrained by time, health, resources,

---

[3] Yunkaporta, *Sand talk.*

trauma, and structural inequality. What matters is not comprehensiveness but alignment – inner steadiness, relational courage, practical skill, civic engagement, and ethical orientation. The world does not require heroic individuals but interconnected communities acting consistently, collectively, and with care.

Finally, even the most determined action cannot prevent all losses. Some systems will fail, some environments will collapse altogether, and some regions will become uninhabitable. Acknowledging these limits does not diminish agency, but clarifies its purpose. Agency at this time is not about preservation of the familiar world but rather shaping, as best we can, the conditions for justice, dignity, and compassion in the world that is emerging.

For celebrated American journalist Dahr Jamail in his groundbreaking book, *The End of Ice: Bearing Witness and Finding Meaning in the Path of Climate Disruption*, this spirit of compassionate curiosity emerges in the most surprising of ways: 'An extraordinary alchemy can take place', he writes, 'when people follow their inner directives to stand up and face squarely the dire odds of biospheric survival. These actions involve extraordinary outer and inner courage, which can nurture a profound activism. The gifts provided by the crisis at hand are the conditions that make possible widespread shifts in policies, identity, purpose and consciousness'.[4] We agree – the climate predicament is our reminder to recalibrate our lives and systems to be more in tune with the deeper truth we have forgotten: that we belong to a vast web of life.

Being in the midst of collapse should, yes, invite us to pause to consider what comes next, but even in a time of darkness and tragedy we can forge new beginnings, new ways of being in the world, with each other and Country.

We invite you to pause, take stock and reflect on what we have shared in these pages, and perhaps to share this book, as well as your own thoughts and feelings with others. Life, for all its travails and unpredictability, is still worth it so long as we realise – beyond the most obvious material necessities – that the things that have always mattered – love, connection, sense of place and Country, meaning and purpose – are the great drivers of the human spirit; the spirit of renewal.

We now invite you to be guided by *your* own inner spark.

---

4 Dahr Jamail, *The end of ice: bearing witness and finding meaning in the path of climate disruption* (New York: The New Press, 2019), 216.

Bibliography

Abebe, Yared A., Amir Ghorbani, Igor Nikolic, Nikola Manojlovic, Andre Gruhn and Zoran Vojinovic, 'The role of household adaptation measures in reducing vulnerability to flooding: a coupled agent-based and flood modelling approach', *Hydrol. earth syst. sci.*, 24 (2020), 5329–54, <https://doi.org/10.5194/hess-24-5329-2020>, accessed 4 October 2025.

Accidental Gods, 'No more business as usual: a second interview with Professor Rupert Read', YouTube video (30 June 2023), <https://www.youtube.com/watch?v=0263sM6L-wQ>, accessed 2 April 2025

Adow, Mohamed, 'The climate debt: what the West owes the rest', *Foreign Affairs* (May/June 2020; published online 13 April 2020), <https://www.foreignaffairs.com/articles/world/2020-04-13/climate-debt>, accessed 7 April 2025.

Akomolafe, Bayo, What climate collapse asks of us (Bayo Akomolafe website, 2019), <https://www.bayoakomolafe.net/post/what-climate-collapse-asks-of-us>, accessed 14 March 2025.

Albom, Mitch, *The five people you meet in heaven* (London: Sphere, 2003).

Aldred, Jessica, 'Australia links 'angry summer' to climate change – at last', *The Guardian* (7 March 2013), <https://www.theguardian.com/environment/blog/2013/mar/07/australia-angry-summer-climate-change>, accessed 6 April 2025.

Aldrich, Daniel P., Building resilience: social capital in post-disaster recovery (Chicago: University of Chicago Press, 2012).

Aldrich, Daniel P., and Michelle A. Meyer, 'Social capital and community resilience', *American Behavioral Scientist* 59/2 (2015), 254–69.

Altieri, Miguel A., and Clara I. Nicholls, 'Agroecology: a brief account of its origins and currents of thought in Latin America', *Agroecology and sustainable food systems*, 41/3–4 (2017), 231–37, <https://doi.org/10.1080/21683565.2017.1287147>, accessed 4 October 2025.

Altieri, Miguel A., and Clara I. Nicholls, 'The adaptation and mitigation potential of traditional agriculture in a changing climate', *Climatic change*, 140 (2017), 33–45, <https://doi.org/10.1007/s10584-013-0909-y>, accessed 4 October 2025.

Amnesty International, The state of the world's human rights: Amnesty International's annual report 2024/25 (29 April 2025), <https://www.amnesty.org/en/latest/news/2025/04/global-human-rights-crisis-trump-effect-accelerates-destructive-trends/>, accessed 12 March 2025.

Anderson, Kevin. 'IPCC's conservative nature masks true scale of action needed to avert catastrophic climate change'. *The Conversation*, 24 March 2023. Available at: <https://theconversation.com/ipccs-conservative-nature-masks-true-scale-of-action-needed-to-avert-catastrophic-climate-change-202287> (accessed 18 March 2025).

Anderson, Kevin, and Glen Peters, 'The trouble with negative emissions', *Science*, 354 (2016), 182–83, <https://doi.org/10.1126/science.aah4567>, accessed 8 August 2025.

Arendt, Hannah, *On violence* (New York: Harcourt, 1970).

Arendt, Hannah, *The origins of totalitarianism* (New York: Harcourt Brace Jovanovich, 1973).

Arriagada, Nerilie B., Palmer, Adrienne J., Bowman, David M. J. S., Morgan, Grant G., Jalaludin, Bikie B. and Johnston, Fay H., 'Unprecedented smoke-related health burden associated with the 2019–20 bushfires in eastern Australia', *Medical Journal of Australia* 213/6 (2020), 282–283.

Australian Academy of Science (AAS), The risks to Australia of a 3°c warmer world (Canberra: Australian Academy of Science, 2021), available at <https://www.science.org.au/supporting-science/science-policy-and-analysis/reports-and-publications/risks-australia-three-degrees-c-warmer-world> (accessed 18 March 2025).

Australian Climate Service, 'National Climate Risk Assessment', <https://www.acs.gov.au/pages/national-climate-risk-assessment>, accessed 26 October 2025.

Australian Institute for Disaster Resilience (AIDR), Communities in post-disaster recovery: working with conflict in and between (Melbourne: AIDR, 2020)

Australian Institute of Health and Welfare (AIHW), *Social cohesion and social connection* (Canberra: AIHW, 2025). <https://www.aihw.gov.au/reports/australias-welfare/social-cohesion-and-social-connection>, accessed 23 October 2025.

Bailey, Sophie, and Nicola Gerrish, 'Ecological grief: how can we bear this together?', *Critical and radical social work*, 13/3 (2025), 348–69, <https://doi.org/10.1332/20498608Y2024D000000059>, accessed 21 November 2025.

Bailie, Jodie, Veronica Matthews, Ross Bailie, Michelle Villeneuve and Joanne Longman, 'Exposure to risk and experiences of river flooding for people with disability and carers in rural Australia: a cross-sectional survey', *BMJ open*, 12/8 (2022), e056210-1–e056210-10.

Baldwin, James, 'As much truth as one can bear: to speak out about the world as it is, says James Baldwin, is the writer's job'. *The New York Times*, 14 January 1962. Available at: <https://www.nytimes.com/1962/01/14/archives/as-much-truth-as-one-can-bear-to-speak-out-about-the-world-as-it-is.html> (accessed 18 March 2025).

Barnett, Jon, and Saffron O'Neill, 'Maladaptation', *Global environmental change*, 20/2 (2010), 211–13, <https://doi.org/10.1016/j.gloenvcha.2009.11.004>, accessed 4 October 2025.

Barlow, Karen, 'Exclusive: government refuses to release climate security report', *The Saturday Paper*, 20 September 2025, available at <https://www.

thesaturdaypaper.com.au/news/politics/2025/09/20/exclusive-government-refuses-release-climate-security-report> (accessed 12 October 2025).

Battistoni, Alyssa, Free gifts: capitalism and the politics of nature (Princeton, NJ: Princeton University Press, 2025).

Bayda, Ezra, *At home in the muddy water: a guide to finding peace within everyday chaos* (Boston and London: Shambhala, 2004).

Bell, Alice, 'Sixty years of climate change warnings: the signs that were missed (and ignored)'. *The Guardian*, 5 July 2021. Available at: <https://www.theguardian.com/science/2021/jul/05/sixty-years-of-climate-change-warnings-the-signs-that-were-missed-and-ignored> (accessed 18 March 2025).

Bendell, Jem, *Breaking together: a freedom-loving response to collapse* (Bristol: Good Works, an imprint of the Schumacher Institute, 2023).

Bendell, Jem, Deep adaptation: a map for navigating climate tragedy (University of Cumbria, 2018), 1–31.

Béné, Christophe, Robert Wood, Andrew Newsham and Mark Davies, *Resilience: new utopia or new tyranny?* IDS Working Paper 405 (Brighton: Institute of Development Studies, 2012).

Beggs, John J., Valerie A. Haines and Jeanne S. Hurlbert, 'The effects of personal network and local community contexts on the receipt of formal aid during disaster recovery', *International journal of mass emergencies & disasters*, 14/1 (1996), 57–78, <https://doi.org/10.1177/028072709601400104>, accessed 4 October 2025.

Bell, Lara, 'Plan C – strengthening communities through free resilience education' (20 August 2023), *The Lismore App*, <https://lismoreapp.com.au/NewsStory/plan-c-strengthening-communities-through-free-resilience-education/64e042d2f5c8a30028e1612b>, accessed 4 October 2025.

Bertana Amanda, Clark, Brett, Benney, Tabitha, and Quackenbush, Cameron, 'Beyond maladaptation: structural barriers to successful adaptation', *Environmental sociology*, 8 (2022).

Bettini, Giovanni, 'Climate barbarians at the gate? the case for a cosmopolitan climate politics'. *Geopolitics* 18/3 (2013). Available at: <https://doi.org/10.1080/14650045.2012.743905> (accessed 18 March 2025).

Binder, Matt, 'Who really profits from climate fear – the billionaires behind the green agenda' (28 January 2025), *Climate Cosmos*, <https://climatecosmos.com/climate-society/who-really-profits-from-climate-fear-the-billionaires-behind-the-green-agenda/>, accessed 17 March 2025.

Blamey, Paul, 'Climate change: media, politicians need to change the rhetoric', *Crikey* (8 April 2024), <https://www.crikey.com.au/2024/04/08/climate-change-rhetoric-media-politics/>, accessed 5 April 2025.

Blockade Australia, Blockade Australia website (no date), <https://www.blockadeaustralia.com/>, accessed 12 March 2025.

Bolton, Matthew, *How to resist: turn protest to power* (London: Bloomsbury Publishing, 2017).

Bowen, Chris, 'Energy and climate change ministerial council press conference', *Department of Climate Change, Energy, the Environment and Water* (7 July 2023), <https://minister.dcceew.gov.au/bowen/transcripts/energy-and-climate-change-ministerial-council-press-conference>, accessed 05 April 2025.

Brach, Tara, *Radical acceptance: embracing your life with the heart of a Buddha* (New York: Bantam, 2003).

Bradley, James, 'Unearthed', *Meanjin* 78/3 (Spring 2019).

Bradshaw, Simon, Lesley Hughes, David Karoly, Joëlle Gergis, Annika Dean, Dinah Arndt, Jennifer Rayner and Jolee Wakefield, *Mission zero: how today's climate choices will reshape Australia*. Climate Council of Australia, 20 September 2023. Available at: <https://climatecouncil.org.au/resources/missionzero/> (accessed 18 March 2025).

Brewer, Judson A., Patrick D. Worhunsky, Jeremy R. Gray, Yi-Yuan Tang, Jochen Weber and Hedy Kober, 'Meditation experience is associated with differences in default mode network activity and connectivity', *Proceedings of the National Academy of Sciences* 108/50 (2011), 20254–59.

Brozović, Danijel, 'Societal collapse: a literature review', *Futures* 145 (2023), article 103075.

Buckton, Sarah, Ioan Fazey, Beth Sharpe et al., 'The regenerative lens: a conceptual framework for regenerative social-ecological systems', *One Earth*, 6/7 (2023), 824–42.

Caldwell, Iain, 'Loneliness epidemic in Australia: Australians finding it harder to make new friends', *The Guardian* (20 September 2025), <https://www.theguardian.com/society/2025/sep/20/loneliness-epidemic-australians-finding-it-harder-to-make-new-friends-hilda-survey>, accessed 15 October 2025.

Camus, Albert, *The myth of Sisyphus*, trans. Justin O'Brien (New York: Vintage International, 1991).

Carmen, Esther, Fazey, Ioan, Ross, Helen, Bedinger, Melissa, Smith, Fiona M., Prager, Katrin, McClymont, Kerri and Morrison, David S., 'Building community resilience in a context of climate change: the role of social capital', *Ambio* 51 (2022), 1371–1387.

Campbell, Richard, Morison, Evan, Ryan, Mark, Saunders, Millie, Le, My N., Adhikari, Ashim, Scicluna, Kade, Simpson, Ellen and Anderson, Lisa, Fossil fuel subsidies in Australia 2024 (13 May 2024, The Australia Institute), <https://australiainstitute.org.au/report/fossil-fuel-subsidies-in-australia-2024/>, accessed 6 April 2025.

Carrington, Damian, 'Climate misinformation turning crisis into catastrophe, report says' (19 June 2025), *The Guardian*, <https://www.theguardian.com/

environment/2025/jun/19/climate-misinformation-turning-crisis-into-catastrophe-ipie-report>, accessed 17 March 2025.

Carrington, Damian. 'World's top climate scientists expect global heating to blast past 1.5c target'. *The Guardian*, 8 May 2024. Available at: <https://www.theguardian.com/environment/2024/may/08/worlds-top-climate-scientists-expect-global-heating-to-blast-past-1-5c-target> (accessed 18 March 2025).

Carter, Peter D., and Woodworth, Elizabeth, Unprecedented crime: climate science denial and game changers for survival (Atlanta, GA: Clarity Press, 2017).

Chancel, Lucas, and Thomas Piketty, *Carbon and inequality: from Kyoto to Paris* (Paris School of Economics, 2015). <http://piketty.pse.ens.fr/files/ChancelPiketty2015.pdf>, accessed 12 May 2025.

Chenoweth, Erica, and Maria J. Stephan, 'Why civil resistance works: the strategic logic of nonviolent conflict', *International Security* 33/1 (2008), 7–44.

Chowdhury, Tazrina, Paul Arbon, Mayumi Kako, Robert Muller, Malinda Steenkamp and Kristine Gebbie, 'Understanding the experiences of women in disasters: lessons for emergency management planning'. *International journal of disaster risk reduction* 104 (2024), 104244. Available at: <https://doi.org/10.1016/j.ijdrr.2023.104244> (accessed 18 March 2025).

Christophers, Brett, The price is wrong: why capitalism won't save the planet (London: Verso Books, 2022).

Climate Action Network Australia (CANA), Dirty power: political donations from the fossil fuel industry (2022, CANA), <https://www.cana.net.au/dirty_power_report>, accessed 6 April 2025.

Climate Analytics, *Rescuing 1.5°c: new evidence on the highest possible ambition to deliver the Paris Agreement*, Berlin, 6 November 2025, available at <https://climateanalytics.org/publications/rescuing-1-5c> (accessed 30 November 2025).

Climate Council, Introducing the dirty dozen: Australia's filthiest fossil fuel polluters (Climate Council, February 2023), <https://www.climatecouncil.org.au/wp-content/uploads/2023/02/Climate-Council-Report-Introducing-the-Dirty-Dozen-Feb-2023-1.pdf>, accessed 7 April 2025.

Climate Council, New analysis: the Albanese government's fossil fuel approvals (3 September 2025, Climate Council), <https://www.climatecouncil.org.au/resources/albanese-governments-fossil-fuel-approvals>, accessed 6 April 2025.

Cohen, Anthony P., *The symbolic construction of community* (London: Tavistock, 1985).

Commonwealth Scientific and Industrial Research Organisation (CSIRO) and Bureau of Meteorology (BOM), *State of the climate 2024* (Commonwealth of Australia, 2024), available at <https://www.csiro.au/

en/research/environmental-impacts/climate-change/State-of-the-Climate/Australias-Changing-Climate> (accessed 18 March 2025).

Comms Declare, *Fossil fuel lobbyists flood Canberra as agencies and media exposed in 2025 F-list* (25 September 2025, Comms Declare), <https://commsdeclare.org/2025/09/25/fossil-fuel-lobbyists-flood-canberra-as-agencies-and-media-exposed-in-2025-f-list/>, accessed 15 October 2025

Cook, John; Nuccitelli, Dana; Green, Sarah A.; Richardson, Mark; Winkler, Bärbel; Painting, Rob; Way, Robert; Jacobs, Peter; Skuce, Andrew. 'Quantifying the consensus on anthropogenic global warming in the scientific literature', *Environmental research letters* 8/2 (2013), 024024.

Copernicus Climate Change Service (C3S), *Climate bulletin*. 8 August 2024. Available at: <https://climate.copernicus.eu/climate-bulletins> (accessed 18 March 2025).

Coutts, Andrew M., Elizabeth C. White, Nigel J. Tapper, Jason Beringer and Stephen J. Livesley, 'Temperature and human thermal comfort effects of street trees across three contrasting street canyon environments', *Theoretical and applied climatology*, 124/1–2 (2016), 55–68, <https://doi.org/10.1007/s00704-015-1409-y>, accessed 4 October 2025.

Creswell, J. David, 'Mindfulness interventions', *Annual Review of Psychology* 68 (2017), 491–516.

Creswell, J. David, Michael R. Irwin, Lisa J. Burklund, Matthew D. Lieberman, Jesus M. G. Arevalo, Julienne Ma, Elizabeth C. Breen and Steve W. Cole, 'Mindfulness-based stress reduction training reduces loneliness and pro-inflammatory gene expression in older adults: a small randomized controlled trial', *Brain, Behavior, and Immunity* 26/7 (2012), 1095–1101.

Cretney, Raven, 'Resilience for whom? emerging critical geographies of socio-ecological resilience', *Geography Compass*, 13/9 (2019), e12440.

Cote, Muriel, and Andrea J. Nightingale, 'Resilience thinking meets social theory: situating social change in socio-ecological systems (SES) research'. *Progress in human geography* 36/4 (2011), 475–89. Available at: <https://doi.org/10.1177/0309132511425708> (accessed 18 March 2025).

Cunsolo, Ashlee, and N. R. Ellis, 'Ecological grief as a mental health response to climate change-related loss', *Nature Climate Change* 8/4 (2018), 275–81.

Curtis, Adam, (dir.), *HyperNormalisation* (BBC, film, 2016).

Cutter, Susan L., 'The forgotten casualties redux: women, children, and disaster risk', *Global environmental change*, 42 (2016), 117–21.

Daly, Herman E., *Ecological economics and sustainable development: selected essays of Herman Daly* (Cheltenham: Edward Elgar Publishing, 2014).

Davidson, Richard J., and Alfred W. Kaszniak, 'Conceptual and methodological issues in research on mindfulness and meditation', *American Psychologist* 70/7 (2015), 581–92.

Davoudi, Simin, Keith Shaw, L. Jamila Haider, Andrea E. Quinlan, Garry D. Peterson, Cathy Wilkinson et al., 'Resilience: a bridging concept or a dead end. the politics of resilience for planning: a cautionary note', *Planning theory & practice*, 13/2 (2012), 299–333, <https://doi.org/10.1080/14649357.2012.677124>, accessed 4 October 2025.

de Oliveira, Vanessa Machado, *Outgrowing Modernity: Navigating Complexity, Complicity, and Collapse with Accountability and Compassion* (Berkeley: North Atlantic Books, 2025).

Delanty, Gerard, *Community*, 2nd edn (London: Routledge, 2010).

Deloitte Access Economics, *South East Queensland rainfall and flooding event – impact assessment*, 2022, commissioned by the Queensland Reconstruction Authority, available at <https://statements.qld.gov.au/statements/95831> (accessed 18 March 2025).

Department of Climate Change, Energy, the Environment and Water (DCCEEW), *National adaptation plan* (Canberra: Commonwealth of Australia, 2025).

Department of Industry, Science and Resources, *Resources and energy major projects: 2024*. Canberra: Commonwealth of Australia, 2024. Available at: <https://www.industry.gov.au/publications/resources-and-energy-major-projects-2024> (accessed 18 March 2025).

Desbordes, Gaëlle, Lobsang T. Negi, Thaddeus W. W. Pace, B. Alan Wallace, Charles L. Raison and Elen N. Schwartz, 'Effects of mindful-attention and compassion meditation training on amygdala response to emotional stimuli in an ordinary, non-meditative state', *Frontiers in Human Neuroscience* 6 (2012), 292.

Di Muzio, Tim, Carbon capitalism: energy, social reproduction and world order (London and New York: Rowman & Littlefield International, 2015).

Dunlop, Ian, 'Climate change and politics as usual – a government missing in action', Ian Dunlop website (7 September 2021), <https://itdunlop.com/climate-change-politics-as-usual-a-government-missing-in-action/>, accessed 5 April 2025.

Dunne, Daisy. 'Analysis: 95% of countries miss UN deadline to submit 2035 climate pledges'. *Carbon Brief*, 10 February 2025. Available at: <https://www.carbonbrief.org/analysis-95-of-countries-miss-un-deadline-to-submit-2035-climate-pledges/> (accessed 18 March 2025).

Dynes, Russell R., *Organized behavior in disaster*, The Disaster Research Center Series (Columbus, Ohio: Disaster Research Center, The Ohio State University, 1974).

Earth.Org, 'New climate report warns of imminent climate tipping point as emissions soar'. Hong Kong, 18 June 2025. Available at: <https://earth.org/new-climate-report-warns-of-imminent-climate-tipping-point-as-emissions-soar/> (accessed 30 November 2025).

Earthhow, '15 climate feedback loops and examples'. 2023. Available at: <https://earthhow.com/climate-feedback-loops/> (accessed 18 March 2025).

EcoMind Australia. Climate science, not climate silence: open letter from scientists and educators. 2025. Available at: <https://www.ecomind.au/open-letter> (accessed 18 March 2025).

Editors of Encyclopaedia Britannica, 'Pakistan floods of 2022', Encyclopaedia Britannica (2024), <https://www.britannica.com/event/Pakistan-floods-of-2022>, accessed 25 March 2025.

Elias, Norbert, *The civilizing process* (Basel: Haus zum Falken, 1939).

Ellerbeck, Stefan, 'Climate change is already altering everything, from fertility choices to insuring our homes', *World Economic Forum* (24 June 2022), <https://www.weforum.org/stories/2022/06/climate-change-weather-extreme-health>, accessed 27 October 2025.

Ensor, Jonathan, and Rachel Berger, Understanding climate change adaptation: lessons from community-based approaches (Practical Action Publishing, 2009), <https://practicalactionpublishing.com/book/2243/understanding-climate-change-adaptation>, accessed 6 October 2025.

Environment Institute of Australia and New Zealand (EIANZ), National adaptation plan issues paper submission (no date), <https://www.eianz.org/document/item/7607>, accessed 4 October 2025.

Eriksen, Christine, and Dale L. Hankins, 'The retention, revival, and subjugation of Indigenous fire knowledge through agency fire fighting in eastern Australia and California', *Society & natural resources*, 27/12 (2014), 1288–1303, <https://doi.org/10.1080/08941920.2014.918226>, accessed 4 October 2025.

Etzioni, Amitai, The spirit of community: rights, responsibilities and the communitarian agenda (New York: Crown Publishers, 1993).

Fairhead, James, Melissa Leach and Ian Scoones, 'Green grabbing: a new appropriation of nature?', *Journal of peasant studies*, 39/2 (2012), 237–61, <https://doi.org/10.1080/03066150.2012.671770>, accessed 8 August 2025.

Farbotko, Caren, and Heather Lazrus, 'The first climate refugees? contesting global narratives of climate change in Tuvalu', *Global environmental change*, 22/2 (2012), 382–90, <https://doi.org/10.1016/j.gloenvcha.2011.11.014>, accessed 8 August 2025.

Featherstone, David, Anthony Ince, Danny Mackinnon, Kendra Strauss and Andrew Cumbers, 'Progressive localism and the construction of political alternatives', *Transactions of the Institute of British Geographers*, 37/2 (2012), 177–82, <http://www.jstor.org/stable/41427938>, accessed 8 August 2025.

Fielding, Victoria, 'Is News Corp following through on its climate change backflip? my analysis of its flood coverage suggests not', *The Conversation* (25 March 2022), <https://theconversation.com/is-news-corp-following-through-on-its-climate-change-backflip-my-analysis-of-its-flood-coverage-suggests-not-179468>, accessed 25 April 2025.

Fisher, Mark, *Capitalist realism: is there no alternative?* (London: Zero Books, 2009).

Folke, Carl, Stephen R. Carpenter, Brian Walker, Marten Scheffer, F. Stuart Chapin III and Johan Rockström, 'Resilience thinking: integrating resilience, adaptability and transformability', *Ecology and Society* 15/4 (2010), 20.

Foote, Wendy, Margaret Alston, David Betts and Tracy McEwan, *Women's leadership and a community 'saving itself': learning from disasters, health and well-being impacts of the Northern Rivers flood 2022*. Open Research Newcastle, 2024. Available at: <https://hdl.handle.net/1959.13/1493441> (accessed 18 March 2025).

Foster, John Bellamy, The ecological rift: capitalism's war on the earth (New York: Monthly Review Press, 2011).

Foster, John, 'Facing up – facing out or facing in? a review of Deep adaptation: navigating the realities of climate chaos, edited by Bendell, J. & Read, R'. (5 July 2021), *Global Discourse Blog*, <https://globaldiscourseblog.co.uk/2021/07/05/facing-up-facing-out-or-facing-in/>, accessed 20 August 2025.

Foucault, Michel, *Fearless speech*, ed. Joseph Pearson (Los Angeles: Semiotext(e), 2001).

Franco, Michael, 'Feedback loops are overlooked engines of climate change, says study', *New Atlas* (17 February 2023), <https://newatlas.com/environment/feedback-loops-climate-change/>, accessed 13 March 2025.

Frankl, Viktor E., *Man's search for meaning* (Boston: Beacon Press, 2006).

Franta, Benjamin, 'Big oil knew: documents reveal fossil fuel industry's decades of climate knowledge', *The Conversation* (25 October 2021), <https://theconversation.com/big-oil-knew-documents-reveal-fossil-fuel-industrys-decades-of-climate-knowledge-170227>, accessed 25 October 2025.

Fraser, Nancy, Cannibal capitalism: how our system is devouring democracy, care, and the planet – and what we can do about it (London: Verso Books, 2022).

Freire, Paulo, *Pedagogy of the oppressed* (New York: Herder and Herder, 1970).

Friedrichs, Jörg, The future is not what it used to be: climate change and energy scarcity. Cambridge, MA: The MIT Press, 2013.

Fukuyama, Francis, *The end of history and the last man* (New York: Free Press, 1992).

Gallagher, James, 'Partners in pollution: Australian universities and the fossil fuel industry', *Redflag* (9 May 2023), <https://redflag.org.au/article/partners-pollution-australian-universities-and-fossil-fuel-industry>, accessed 6 April 2025.

Gardiner, Stephen M., A perfect moral storm: the ethical tragedy of climate change (Oxford: Oxford University Press, 2011).

Gergis, Joëlle. 'Highway to hell: climate change and Australia's future'. *The Australia Institute*, 28 June 2024. Available at: <https://australiainstitute.org.au/event/highway_to_hell/> (accessed 18 March 2025).

Gergis, Joëlle, *Humanity's moment: a climate scientist's case for hope* (Melbourne: Black Inc., 2022).

Gergis, Joëlle. 'Friday essay: "I feel my heart breaking today" – a climate scientist's path through grief towards hope'. *The Conversation*, 26 August 2022. Available at: <https://theconversation.com/friday-essay-i-feel-my-heart-breaking-today-a-climate-scientists-path-through-grief-towards-hope-188704> (accessed 18 March 2025).

Gergis, Joëlle, 'Net zero distracts from scientific imperative to eliminate fossil fuels', *The Guardian* (7 November 2025), <https://www.theguardian.com/commentisfree/2025/nov/07/net-zero-distracts-from-scientific-imperative-to-eliminate-fossil-fuels-cop30>, accessed 25 November 2025.

Gergis, Joëlle, Grant Blashki, Jane Gardner and Simon Bradshaw, Climate trauma: the growing toll of climate change on the mental health of Australians. Sydney: Climate Council of Australia, 2023. Available at: <https://www.climatecouncil.org.au/wp-content/uploads/2023/02/Report-Climate-Change-and-Mental-Health.pdf> (accessed 27 October 2025).

Girardet, Herbert, *Creating regenerative cities* (Oxon: Routledge, 2014).

Gramling, Carolyn, 'Coral collapse signals Earth's first climate tipping point'. *Science News*, Washington DC, 12 October 2025. Available at: <https://www.sciencenews.org/article/coral-collapse-climate-tipping-point> (accessed 30 November 2025).

Granovetter, Mark, 'The strength of weak ties', *American journal of sociology*, 78/6 (1973), 1360–80.

Greer, John, Michael, The Long Descent: A User's Guide to the End of the Industrial Age (Canada: New Society Publishers 2008)

Griscom, Bronson W., et al., 'Natural climate solutions', *Proceedings of the National Academy of Sciences*, 114/44 (2017), 11645–50, <https://doi.org/10.1073/pnas.1710465114>, accessed 8 August 2025.

Goodell, Jeff, *The heat will kill you first: life and death on a scorched planet* (New York: Little, Brown and Company, 2023).

Gorissen, Leen, interview with wearethe_regeneration, Instagram post (18 January 2026).

Gosnell, Hannah, Nicholas Gill and Michelle Voyer, 'Transformational adaptation on the farm: processes of change and persistence in transitions to "climate-smart" regenerative agriculture', *Global environmental change*, 59 (2019), article 101965, <https://doi.org/10.1016/j.gloenvcha.2019.101965>, accessed 4 October 2025.

Goyal, Madhav, Sonal Singh, Erica M. S. Sibinga, Nancy F. Gould, Ashley Rowland-Seymour, Ritu Sharma, Zachary Berger, David Sleicher, Daniel D. Maron, Husam M. Shihab, Priyanga D. Ranasinghe, Steven Linn,

Somnath Saha, Eric B. Bass and Jennifer A. Haythornthwaite, 'Meditation programs for psychological stress and well-being: a systematic review and meta-analysis', *JAMA Internal Medicine* 174/3 (2014), 357–68.

Graham, Stephen, 'Svante Arrhenius', NASA Earth Observatory (18 January 2000), <https://earthobservatory.nasa.gov/features/Arrhenius>, accessed 12 March 2025.

Gramsci, Antonio, *Selections from the prison notebooks*, ed. and trans. Quintin Hoare and Geoffrey Nowell Smith (London: Lawrence & Wishart, 1971).

Great Barrier Reef Foundation (GBRF), 'Climate change is the single biggest threat facing the Reef', 2025, available at <https://www.barrierreef.org/the-reef/threats/climate-change> (accessed 18 March 2025).

Hall, Shannon, 'Exxon knew about climate change almost 40 years ago'. *Scientific American*, 26 October 2015. Available at: <https://www.scientificamerican.com/article/exxon-knew-about-climate-change-almost-40-years-ago/> (accessed 6 November 2025).

Hamilton, Clive, *Climate change and social collapse*, Research paper 7, Climate Adaptation Project, Charles Sturt University, December 2025, p. 1.

Hansen, James E., Pushker Kharecha, Makiko Sato, George Tselioudis, Jon Kelly, Susanne E. Bauer … Auli Pokela. 'Global warming has accelerated: are the United Nations and the public well-informed?' Environment: science and policy for sustainable development 67/1 (2025), 6–44. Available at: <https://doi.org/10.1080/00139157.2025.2434494> (accessed 18 March 2025).

Hansen, James E., Sato, Makiko, Simons, Leon, Nazarenko, Larissa S., Sangha, Inderpreet, Kharecha, Pushker, Zachos, James C., von Schuckmann, Karina, Loeb, Norman G., Osman, Marlene B., Jin, Qinjian, Tselioudis, Giorgi, Jeong, Eunjee, Lacis, Andrew, Ruedy, Reto, Russell, Gary, Cao, Jingyi and Li, Jiang, 'Global warming in the pipeline', *Oxford Open Climate Change* 3/1 (2023).

Haraway, Donna J., *Staying with the trouble: making kin in the Chthulucene* (Durham and London: Duke University Press, 2016).

Hari, Johann, *Lost connections: uncovering the real causes of depression – and the unexpected solutions* (London: Bloomsbury, 2018).

Hartman, Chester, and Squires, Gregory D., eds, *There Is No Such Thing as a Natural Disaster: Race, Class, and Hurricane Katrina* (New York: Routledge, 2006).

Head, Brian W., 'Wicked Problems in Public Policy', *Public Policy* 3/2 (2008), 101–118.

Head, Lesley, Kindred by nature: grief and courage in a climate-changed world (Sydney: UNSW Press, 2022).

Hémon, Daniel, and Jougla, Éric, 'Surveillance of heat-related mortality in France during the summer of 2003', *Eurosurveillance* 9/8 (2004), 6–8.

Herzog, Jonathan P., *The spiritual-industrial complex: America's religious battle against communism in the early Cold War* (New York: Oxford University Press, 2011).

Hiatt, Juanima, 'A resilient heart: navigating life's storms with strength' (21 October 2023), <https://medium.com/illumination/a-resilient-heart-navigating-lifes-storms-with-strength-9644a42d3fa3>, accessed 8 August 2025.

Hickel, Jason, *Less is more: how degrowth will save the world* (London: William Heinemann, 2020).

Hickel, Jason, and Giorgos Kallis, 'Is green growth possible?', New political economy, 25/4 (2020), 469–86, <https://doi.org/10.1080/13563467.2019.1598964>, accessed 4 October 2025.

Hickman, Caroline, Elizabeth Marks, Panu Pihkala, Susan Clayton, Rebecca E. Lewandowski, Emily E. Mayall, Britt Wray, Christopher Mellor and Lise van Susteren, 'Climate anxiety in children and young people and their beliefs about government responses to climate change: a global survey', *The Lancet Planetary Health* 5/12 (2021), e863–e873.

Hil, Richard, Louise Holdsworth and Charlie Brennan, *Being lonely* (Ngara Institute and Resilient Byron, 2020). <https://www.planc.org.au/beinglonely>, accessed 7 April 2025.

Hill, Ruth, Fiona J. Walsh, Jocelyn Davies, Annie Sparrow, Moira Mooney, Russell M. Wise and Maria Tengö, 'Knowledge co-production for Indigenous adaptation pathways: transformative learning, action, and renewal', *Ecology and society*, 17/3 (2013), 12.

Hilton, Doug, 'We need science, and we need to defend it from attack' (18 April 2024), *CSIRO News*, <https://www.csiro.au/en/news/all/articles/2024/april/defending-science>, accessed 17 March 2025.

Hine, Dougald, At work in the ruins: finding our place in the time of science, climate change, pandemics and all the other emergencies (Vermont: Chelsea Green Publishing, 2023).

Hino, Miyuki, Christopher Field and Katharine Mach, 'Managed retreat as a response to natural hazard risk', *Nature climate change*, 7 (2017), 364–70, <https://doi.org/10.1038/nclimate3252>, accessed 4 October 2025.

Hoffman, Andrew J., *How Culture Shapes the Climate Change Debate* (Stanford, CA: Stanford University Press, 2015).

Hoegh-Guldberg, Ove, and Lesley Hughes, 'Seriously ugly: here's how Australia will look if the world heats by 3°c this century'. *The Conversation*, 31 March 2021. Available at: <https://theconversation.com/seriously-ugly-heres-how-australia-will-look-if-the-world-heats-by-3-c-this-century-157875> (accessed 18 March 2025).

Hoge, Elizabeth A., Eric Bui, Sarah A. Palitz, Nicole R. Schwarz, Margaret E. Owens, James M. Johnston, Mark H. Pollack, Naomi M. Simon and Scott P. Orr, 'Mindfulness-based stress reduction vs escitalopram for the treatment of adults with anxiety disorders: a randomized clinical trial', *JAMA Psychiatry* 79/1 (2022), 13–21.

Holling, Crawford S., 'Resilience and stability of ecological systems', *Annual review of ecology and systematics*, 4 (1973), 1–23, <https://doi.org/10.1146/annurev.es.04.110173.000245>, accessed 4 October 2025.

Holmgren, David, *Permaculture: principles and pathways beyond sustainability* (Hepburn: Holmgren Design Services, 2002).

Huber, Benedikt, 'Countries spend huge sums on fossil fuel subsidies – why they're so hard to eliminate', *The Conversation* (13 November 2024), <https://theconversation.com/countries-spend-huge-sums-on-fossil-fuel-subsidies-why-theyre-so-hard-to-eliminate-239346>, accessed 6 April 2025.

Hughes, Lesley, James Rayner, Andrew Dean and Ketan Tidswell, *Dangerously overlooked: why we need to talk about methane*. Sydney: Climate Council, 2024. Available at: <https://www.climatecouncil.org.au/resources/dangerously-overlooked-why-we-need-to-talk-about-methane/> (accessed 18 March 2025).

Hülsen, Sebastian, Laura E. Dee, Christian M. Kropf et al., 'Mangroves and their services are at risk from tropical cyclones and sea level rise under climate change', *Commun earth environ*, 6 (2025), article 262, <https://doi.org/10.1038/s43247-025-02242-z>, accessed 4 October 2025.

Hurlbert, Jeanne S., Valerie A. Haines and John J. Beggs, 'Core networks and tie activation: what kinds of routine networks allocate resources in nonroutine situations?', *American sociological review*, 65/4 (2000), 598–618, <https://doi.org/10.2307/2657385>, accessed 4 October 2025.

Huq, Saleemul, 'Improving information for community-based adaptation' (2011), IIED Opinion Paper, <https://www.iied.org/17103iied>, accessed 4 October 2025.

Huq, Saleemul, Hannah Reid, Mozaharul Alam, Rachel Berger, Terry Cannon and Andrew Milligan, Community-based adaptation to climate change: an overview (PLA 60), *Participatory learning and action*, 60 (2009), <https://www.iied.org/g02608>, accessed 4 October 2025.

Ignatieff, Michael, 'The art of consolation: how to find solace in dark times', *Persuasion* (24 November 2022). <https://www.persuasion.community/p/the-art-of-consolation>, accessed 12 October 2025.

Insurance Council of Australia (ICA), 'Catastrophe report: cyclone Alfred – March 2025', 2025, available at <https://insurancecouncil.com.au/reports/cyclone-alfred-catastrophe-report-march-2025> (accessed 30 March 2025).

Intergovernmental Panel on Climate Change (IPCC), Climate change 2022: impacts, adaptation, and vulnerability. Contribution of Working Group II to the Sixth Assessment Report of the Intergovernmental Panel on Climate Change, eds Hans-Otto Pörtner, Debra C. Roberts, Marina Tignor, Elvira S. Poloczanska, Katja Mintenbeck, Alegría A. Alegría, Melinda Craig, Sarah

Langsdorf, Sonja Löschke, Viola Möller, Aderinto Okem and Babatunde Rama (Cambridge: Cambridge University Press, 2022), <https://www.ipcc.ch/report/ar6/wg2/>, accessed 25 November 2025.

Intergovernmental Panel on Climate Change (IPCC), *Climate change 2023: synthesis report. contribution of working groups I, II and III to the sixth assessment report of the Intergovernmental Panel on Climate Change* (Geneva: IPCC, 2023).

Intergovernmental Panel on Climate Change (IPCC), *Sixth Assessment Report* (no date), <https://www.ipcc.ch/assessment-report/ar6/>, accessed 4 October 2025.

International Institute for Environment and Development (IIED), 'Shifting climate adaptation finance to local communities: evidence and lessons' (2025), Briefing Paper, International Institute for Environment and Development, London, <https://www.iied.org/22654iied>, accessed 22 October 2025.

International Energy Agency (IEA), *World energy investment 2025 – United States*. 2025. Available at: <https://www.iea.org/reports/world-energy-investment-2025/united-states> (accessed 18 March 2025).

International Institute for Sustainable Development (IISD), *Carbon minefields: oil and gas exploration monitor – July edition*. 2024. Available at: <https://www.iisd.org/publications/newsletter/carbon-minefields-monthly-oil-gas-exploration-monitor-july-2024> (accessed 18 March 2025).

International Union for Conservation of Nature (IUCN), Financing ecosystem-based adaptation: trends, barriers, and opportunities (2024), available at <https://iucn.org/blog/202511/financing-ecosystem-based-adaptation-trends-barriers-and-opportunities> (accessed 18 March 2025).

Ivanova, Diana, John Barrett, Dominik Wiedenhofer, Biljana Macura, Max W. Callaghan and Felix Creutzig, 'Quantifying the potential for climate change mitigation of consumption options', *Environmental research letters* (2020), <https://dx.doi.org/10.1088/1748-9326/ab8589>, accessed 8 August 2025.

Iyke, Blessing, N., 'For decades, governments have subsidised fossil fuels. But why?', *The Conversation* (30 July 2024), <https://theconversation.com/for-decades-governments-have-subsidised-fossil-fuels-but-why-213467>, accessed 6 April 2025.

Jackson, Tim, *Post growth: life after capitalism* (Cambridge: Polity Press, 2021).

Jeans, Helen, Gonzalo E. Castillo and Sarah Thomas, *Absorb, adapt, transform: resilience capacities* (Oxfam, 2017).

Jamail, Dahr, The end of ice: bearing witness and finding meaning in the path of climate disruption (New York: The New Press, 2019).

Jenkinson, Stephen, *Die wise: a manifesto for sanity and soul* (Berkeley, California: North Atlantic Books, 2015).

Jordan, Martin, *Nature and therapy: understanding counselling and psychotherapy in outdoor spaces* (London: Routledge, 2009).

Joseph, Jonathan, 'Resilience as embedded neoliberalism: a governmentality approach'. *Resilience* 1/1 (2013), 38–52. Available at: <https://doi.org/10.1080/21693293.2013.765741> (accessed 18 March 2025).

Joshi, Ketan, 'Climate fatalism is just as wrong as climate denial', *Crikey* (14 July 2025), <https://www.crikey.com.au/2025/07/14/climate-fatalism-as-bad-as-climate-denial>, accessed 12 October 2025.

Joshi, Ketan, 'The safeguard mechanism's pro-fossil flaws – explained' (11 August 2025), <https://australiainstitute.org.au/post/the-safeguard-mechanisms-pro-fossil-flaws-explained>, accessed 8 August 2025.

Kabat-Zinn, Jon, *Coming to our senses: healing ourselves and the world through mindfulness* (New York: Hyperion, 2005).

Kabat-Zinn, Jon, Full catastrophe living: Using the wisdom of your body and mind to face stress, pain, and illness (New York: Delacorte, 1990).

Kahan, Dan M., 'Why we are poles apart on climate change', *Nature* 488 (2012), 255.

Kallis, Giorgos, *Limits: why Malthus was wrong and why environmentalists should care* (Stanford: Stanford University Press, 2019).

Kaniasty, Krzysztof, and Fran H. Norris, 'A test of the social support deterioration model in the context of natural disaster', *Journal of personality and social psychology*, 64/3 (1993), 395–408, <https://doi.org/10.1037/0022-3514.64.3.395>, accessed 4 October 2025.

Kassinove, Howard, and Raymond C. Tafrate, *Anger management: the complete treatment guidebook for practitioners* (Atascadero, California: Impact Publishers, 2002).

Kaza, Stephanie, Green Buddhism: practice and compassionate action in uncertain times (Boulder, Colorado: Shambhala Publications, 2019).

Keane, Bernard, 'Election '25: two campaigns and absolutely no winners', Crikey (2 May 2025), <https://www.crikey.com.au/2025/05/02/2025-federal-election-anthony-albanese-peter-dutton-campaigns/>, accessed 5 April 2025.

Keane, Bernard, 'Tanya Plibersek approves Santos gas project', *Crikey* (22 February 2023), <https://www.crikey.com.au/2023/02/22/tanya-plibersek-approves-santos-gas-project/>, accessed 6 April 2025.

Keane, Dean, 'Protest penalties, including jail time and $50k fines, prompt backlash from human rights groups', *ABC News* (19 May 2023), <https://www.abc.net.au/news/2023-05-19/unions-and-human-rights-groups-slam-sa-protest-law-change/102367076>, accessed 12 March 2025.

Kearns, Jonathan, Mike Major and David Norman, 'Why is Australian household debt high relative to history and other countries?'. *Research Discussion Paper* 2020–05. Reserve Bank of Australia (RBA), 2020. Available at: <https://www.rba.gov.au/publications/rdp/2020/2020-05/why-is-australian-household-debt-high-relative-to-history-and-other-countries.html> (accessed 18 March 2025).

Kellert, Stephen R., and Edward O. Wilson, eds, *The biophilia hypothesis* (Washington, DC: Island Press, 1993).

Kemp, Luke, 'Are we on the road to civilisation collapse?', *BBC Future* (19 January 2019), <https://www.bbc.com/future/article/20190218-are-we-on-the-road-to-civilisation-collapse>, accessed 12 March 2025.

Kemp, Luke, *Goliath's curse: the history and future of societal collapse* (New York: Viking, 2025).

Kenis, Annelies, and Matthias Lievens, The limits of the green economy: from re-inventing capitalism to re-politicising the present (London and New York: Routledge, 2014).

Khoury, Bassam, Tania Lecomte, Guillaume Fortin, Monique Massé, Philippe Thérien, Vincent Bouchard, Marie-Andrée Chapleau, Karine Paquin and Stefan G. Hofmann, 'Mindfulness-based therapy: a comprehensive meta-analysis', *Clinical Psychology Review* 33/6 (2013), 763–71.

Kierkegaard, Søren, *The sickness unto death: a Christian psychological exposition for upbuilding and awakening*, trans. Howard V. Hong and Edna H. Hong (Princeton, New Jersey: Princeton University Press, 1980).

Kimmerer, Robin Wall, *Braiding sweetgrass: indigenous wisdom, scientific knowledge, and the teachings of plants* (Minneapolis, MN: Milkweed Editions, 2013).

Kingsnorth, Paul, and Hine, Dougald, *Uncivilisation: the dark mountain manifesto* (online publication, 2009), <https://dark-mountain.net/about/manifesto>, accessed 28 October 2025.

Klein, Naomi and Taylor, Astra, 'The rise of end times fascism', *The Guardian* (13 April 2025), <https://www.theguardian.com/us-news/ng-interactive/2025/apr/13/end-times-fascism-far-right-trump-musk>, accessed 12 October 2055.

Klein, Naomi, The shock doctrine: the rise of disaster capitalism (New York: Metropolitan Books, 2007).

Klein, Naomi, *This Changes Everything: Capitalism vs. the Climate* (London: Penguin Books, 2014).

Kleres, Jochen, and Åsa Wettergren, 'Fear, hope, anger, and guilt in climate activism', *Social Movement Studies* 16/5 (2017), 507–19.

Klinenberg, Eric, Heat wave: a social autopsy of disaster in Chicago (Chicago: University of Chicago Press, 2002).

Kretzmann, John P., and John L. McKnight, Building communities from the inside out: a path toward finding and mobilizing a community's assets (Evanston: Center for Urban Affairs and Policy Research, Northwestern University, 1993).

Krishnamurti Foundation Trust (KFT), 'Regarding the quote "It is no measure of health…"'. [No date]. Available at: <https://kfoundation.org/it-is-no-measure-of-health-to-be-well-adjusted-to-a-profoundly-sick-society/> (accessed 18 March 2025).

Krugman, Paul, *The great unravelling: losing our way in the new century* (New York: W. W. Norton & Company, 2003).

Lakoff, George, 'Why it Matters How We Frame the Environment', Environmental Communication: A Journal of Nature and Culture 4/1 (2010), 70–81.

Lane, David, Global neoliberal capitalism and the alternatives: from social democracy to state capitalisms (Bristol: Policy Press, 2023).

Lederach, John Paul, *The moral imagination: The art and soul of building peace* (Oxford and New York: Oxford University Press, 2005).

Lee, Wendy L., Eco-nihilism: the philosophical geopolitics of the climate change apocalypse (Lanham, Pennsylvania: Lexington Books, 2017).

Leiserowitz, Anthony, Edward Maibach, Seth Rosenthal, John Kotcher, Jennifer Carman, Lisa Neyens, Teresa Myers, Matthew Goldberg, Emily Campbell, Karine Lacroix and Jennifer Marlon, *Climate change in the American mind, April 2022* (New Haven, CT: Yale Program on Climate Change Communication, Yale University and George Mason University, 2022).

Lent, Jeremy, 'Steven Pinker's ideas are fatally flawed. These eight graphs show why', *Open Democracy* (21 May 2018), <https://www.opendemocracy.net/en/transformation/steven-pinker-s-ideas-are-fatally-flawed-these-eight-graphs-show-why/>, accessed 12 March 2025.

Levermann, Anders, 'Individuals can't solve the climate crisis. governments need to step up' (10 July 2019), <https://www.theguardian.com/commentisfree/2019/jul/10/individuals-climate-crisis-government-planet-priority>, accessed 8 August 2025.

Lewis, Sarah, *The rise: creativity, the gift of failure, and the search for mastery* (London: William Collins, 2014).

Lewis, Vivian Marie, 'Learning from Bertrand Russell in today's tumultuous world' (14 May 2018), *The Conversation*, <https://theconversation.com/learning-from-bertrand-russell-in-todays-tumultuous-world-95405>, accessed 4 October 2025.

Linehan, Marsha M., *Cognitive-behavioural treatment of borderline personality disorder* (New York: Guilford Press, 1993).

Logan, Tyne. 'In 2024 global warming topped 1.5c: have we failed the Paris Agreement?' *ABC News*, 13 January 2024. Available at: <https://www.abc.net.au/news/2025-01-13/a-year-above-1-5c-what-does-it-really-mean/104774364> (accessed 18 March 2025).

Lorde, Audre, 'The uses of anger: women responding to racism', in Cherríe Moraga and Gloria Anzaldúa, eds, *This bridge called my back: writings by radical women of color* (Watertown, Massachusetts: Persephone Press, 1981), 127–33.

Loy, David R., Ecodharma: Buddhist teachings for the ecological crisis (Boston: Wisdom Publications, 2019).

Lynas, Mark, Benjamin Z. Houlton and Simon Perry, 'Greater than 99% consensus on human caused climate change in the peer-reviewed scientific literature', *Environmental research letters* 16/11 (2021), 114005. Available at: <https://iopscience.iop.org/article/10.1088/1748-9326/ac2966> (accessed 18 March 2025).

MacCarthy, James, Jessica Richter, Sasha Tyukavina and Nancy Harris, 'The latest data confirms: forest fires are getting worse'. *World Resources Institute*, 21 July 2025. Available at: <https://www.wri.org/insights/global-trends-forest-fires> (accessed 18 March 2025).

MacKintosh, Rachael, 'A climate adaptation model for Māori groups, Aotearoa New Zealand', *Te kaharoa*, 18/1 (2025), 45–68, <https://doi.org/10.24135/tekaharoa.v18i1.483>, accessed 4 October 2025.

Macy, Joanna, 'The Greatest Danger', *YES! Magazine* (2 February 2008), <https://www.yesmagazine.org/issue/climate-solutions/opinion/2008/02/02/the-greatest-danger>, accessed 26 October 2025.

Macy, Joanna, and Johnstone, Chris, Active hope: how to face the mess we're in with unexpected resilience and creative power (revised edn, Novato, CA: New World Library, 2022).

Macy, Joanna, 'Working through environmental despair', in Theodore Roszak, Mary E. Gomes and Allen D. Kanner (eds.), *Ecopsychology* (San Francisco: Sierra Club, 1995; page numbers not supplied).

Maguire, Bill, 'Opinion: I'm a climate scientist. if you knew what I know, you'd be terrified too' (7 March 2024), *CNN Opinion*, <https://edition.cnn.com/2024/03/07/opinions/climate-scientist-scare-doom-anxiety-mcguire>, accessed 14 July 2025.

Maharaj, Nisargadatta, *I am that*, trans. Maurice Frydman (Durham, North Carolina: Acorn Press, 1999) (original work published 1973).

Maharshi, Ramana, *Talks with Sri Ramana Maharshi*, 2nd edn (Carlsbad, California: Inner Directions, 2000) (original work published 1955).

Maier, Steven F., and Martin E. P. Seligman, 'Learned helplessness at fifty: insights from neuroscience', *Psychological Review* 123/4 (2016), 349–67.

Malinas, Mikayla, Biggins, Hannah, Hall, Tessa, Banyasz, Daisy and Paras, Varun, 'Activist shareholders are becoming more efficient, more sophisticated and better resourced', *Allens* (23 September 2024), <https://www.allens.com.au/insights-news/insights/2024/09/activist-shareholders-are-becoming-more-efficient-more-sophisticated-and-better-resourced/>, accessed 12 March 2025.

Mang, Pam, and Bill Reed, 'Designing from place: a regenerative framework and methodology', *Building research & information*, 40/1 (2012), 23–38.

Mang, Pam, and Bill Reed, *Regenerative development & design: a framework for evolving sustainability* (New Jersey: Wiley, 2016).

Mann, Michael, and Peter Hotez, 'Science is under siege from weaponised disinformation – posing a threat to human civilisation' (9 September 2025), *Australian Online News*, <https://australianonlinenews.com.au/2025/09/09/science-is-under-siege-from-weaponised-disinformation-posing-a-threat-to-human-civilisation-michael-mann-and-peter-hotez/>, accessed 4 October 2025.

Manyena, Siambabala, B., 'The concept of resilience revisited', *Disasters*, 30/4 (2006), 433–50, <https://doi.org/10.1111/j.0361-3666.2006.00331.x>, accessed 4 October 2025.

Markusson, Nils, Duncan McLaren, Bronislaw Szerszynski et al., 'Life in the hole: practices and emotions in the cultural political economy of mitigation deterrence', *European journal of futures research*, 10 (2022), article 2, <https://doi.org/10.1186/s40309-021-00186-z>, accessed 8 August 2025.

Markusson, Nils, Duncan McLaren and David Tyfield, 'Towards a cultural political economy of mitigation deterrence by negative emissions technologies (NETs)', *Global sustainability*, 1/e10 (2018), 1–9, <https://doi.org/10.1017/sus.2018.10>, accessed 8 August 2025.

Market Forces, *Banking climate failure: big four Australian banks pour $3.6 billion into fossil fuels in 2023* (16 July 2024, Market Forces), <https://www.marketforces.org.au/media-release-banking-climate-failure-2024/>, accessed 6 April 2025.

Market Forces, Dirty and dark: political donations from the fossil fuel industry (1 February 2023, Market Forces), <https://www.marketforces.org.au/politicaldonations2023/>, accessed 6 April 2025.

Market Forces, Super funds invest $39 billion in fossil fuel growth (28 May 2024, Market Forces), <https://www.marketforces.org.au/super-funds-invest-39-billion-in-fossil-fuel-growth/>, accessed 6 April 2025.

Markowitz, Ezra M., and Guckian, Meaghan L., 'Climate change communication: Challenges, insights, and opportunities', in Clayton, Susan and Manning, Christie (eds), *Psychology and Climate Change: Human Perceptions, Impacts, and Responses* (Oxford: Academic Press, 2018), 35–63.

Martela, Frank, and Michael F. Steger, 'The three meanings of meaning in life: distinguishing coherence, purpose, and significance', *The Journal of Positive Psychology* 11/5 (2016), 531–45.

Maté, Gabor, and Daniel Maté. *The myth of normal: trauma, illness, and healing in a toxic culture*. New York: Avery, 2022.

Mathiesen, Karl, Kalina Oroschakoff, Giorgio Coi and Aitor Busquets Guàrdia, 'Droughts, fires and floods: how climate change will impact Europe'. *POLITICO Europe*, 2021. Available at: <https://www.politico.eu/article/how-climate-change-will-widen-european-divide-road-to-cop26/> (accessed 18 March 2025).

Mazengarb, Michael, 'Millions in fossil fuel money rained on major political parties last year', *Renew Economy* (1 February 2024), <https://reneweconomy.com.au/millions-in-fossil-fuel-money-rained-on-major-political-parties-last-year/>, accessed 6 April 2025.

McCann, Lisa A., and Laurie Anne Pearlman. 'Vicarious traumatisation: a framework for understanding the psychological effects of working with victims'. *Journal of traumatic stress* 3/1 (1990): 131–49. Available at: <https://link.springer.com/article/10.1007/BF00975140> (accessed 18 March 2025).

McIlroy, Tom, 'Labor's 2035 emissions target may be middle of the road but it shows there's no going back on climate', *The Guardian* (19 September 2025), <https://www.theguardian.com/environment/2025/sep/19/labors-2035-emissions-target-may-be-middle-of-the-road-but-it-shows-theres-no-going-back-on-climate>, accessed 6 April 2025.

McKibben, Bill, 'Capitalism is acting truly suicidally on climate', *Common Dreams* (18 April 2024), <https://www.commondreams.org/opinion/capitalism-suicidal-climate>, accessed 7 April 2025.

McLennan, Jim, and Adrian Birch, 'A potential crisis in wildfire emergency response capability? Australia's volunteer firefighters', *Global environmental change part B: environmental hazards*, 6/2 (2005), 101–7, <https://doi.org/10.1016/j.hazards.2005.10.003>, accessed 8 August 2025.

McLisky, Claire, and Emma Pittaway, Case study: Australia – how mutual aid strengthened Greater Uki's community resilience, 2019–2022 (March 2025), with contributions by Jean Renouf, Jo Longman and Richard Hil, under the direction of Pablo Servigne and Véronique de Geoffroy (Plan C and Groupe URD, 2025). <https://www.planc.org.au/mutual-aid>, accessed 26 November 2025.

McMillan, David W., and David M. Chavis, 'Sense of community: a definition and theory', *Journal of Community Psychology* 14/1 (1986), 6–23.

Mecking, Olga, *Niksen: Embracing the Dutch art of doing nothing* (London: Little, Brown Book Group, 2020).

Melbourne Institute, The Household, Income and Labour Dynamics in Australia (HILDA) survey: selected findings from waves 1 to 21 (Melbourne: Melbourne Institute, The University of Melbourne, 2025). <https://melbourneinstitute.unimelb.edu.au/hilda>, accessed 23 October 2025.

Mills, Ciaran, Prepared: how to survive in the real world and find the way forward in the crisis (London: Penguin, 2019).

Milman, Oliver, 'Joe Rogan claims study shows Earth cooling – but report's authors say he's wrong', *The Guardian* (4 September 2025), <https://www.theguardian.com/environment/2025/sep/04/joe-rogan-climate-coolin g-misinformation>, accessed 05 September 2025.

Milman, Oliver, and Damian Carrington. 'World still on track for catastrophic 2.6c temperature rise, report finds'. *The Guardian*, 13 November 2025. Available at: <https://www.theguardian.com/environment/2025/nov/13/worl d-still-on-track-catastrophic-2-6c-temperature-rise-report-finds> (accessed 18 March 2025).

Milman, Oliver, and Nina Lakhani. ‘Wealthy western countries lead in global oil and gas expansion’. *The Guardian*, 25 July 2024. Available at: <https://www.theguardian.com/environment/article/2024/jul/24/new-oil-gas-emission-data-us-uk> (accessed 18 March 2025).

Mocatta, Grahame, and Hawley, Emily, ‘Uncovering a climate catastrophe? media coverage of Australia’s Black Summer bushfires and the revelatory extent of the climate blame frame’, *M/C Journal* 23/4 (2020), <https://doi.org/10.5204/mcj.1666>.

Mol, Arthur P. J., and Spaargaren, Gert, ‘Ecological modernisation theory in debate: a review’, Environmental Politics 9/1 (2000), 17–49.

Mollison, Bill, *Permaculture: a designer’s manual* (Tyalgum, New South Wales: Tagari Publications, 1988).

Mombauer, Dennis, Ann-Christine Link and Kees van der Geest, ‘Addressing climate-related human mobility through NDCs and NAPs: state of play, good practices, and the ways forward’, *Frontiers in climate*, 5 (14 March 2023), article 1125936, <https://doi.org/10.3389/fclim.2023.1125936>, accessed 4 October 2025.

Monash Centre for Youth Policy and Education Practice, ‘Australian Youth Barometer 2024’, *Monash University, Lens* (1 August 2024), <https://lens.monash.edu/@education/2024/08/01/1386898/youth-barometer-paints-a-bleak-picture-of-young-australians-in-2024>, accessed 12 March 2025.

Monbiot, George, ‘Capitalism is killing the planet – it’s time to stop buying into our own destruction’ (30 October 2021), <https://www.theguardian.com/environment/2021/oct/30/capitalism-is-killing-the-planet-its-time-to-stop-buying-into-our-own-destruction>, accessed 8 August 2025.

Monbiot, George, ‘The big polluters’ masterstroke was to blame the climate crisis on you and me’, *The Guardian* (9 October 2019), <https://www.theguardian.com/commentisfree/2019/oct/09/polluters-climate-crisis-fossil-fuel>, accessed 25 November 2025.

Morison, Evan, *et al.*, Climate of the nation 2024: tracking Australia’s attitudes towards climate change and energy (Canberra: The Australia Institute, 2024).

Morris, Hannah, Apocalyptic authoritarianism: climate crisis, media, and power (Oxford: Oxford University Press, 2025).

Mortlock, Timothy, ‘How rising sea levels will affect our coastal cities and towns’. *Australian Geographic*, 24 January 2024. Available at: <https://www.australiangeographic.com.au/news/2024/01/how-rising-sea-levels-will-affect-our-coastal-cities-and-towns/> (accessed 18 March 2025).

Morton, Timothy, *Hyperobjects: Philosophy and Ecology after the End of the World* (Minneapolis: University of Minnesota Press, 2013).

Moses, Asher, 'Collapse of civilisation is the most likely outcome: top climate scientists', *Resilience* (8 June 2020), <https://www.resilience.org/stories/2020-06-08/collapse-of-civilisation-is-the-most-likely-outcome-top-climate-scientists/>, accessed 15 October 2025.

Muller, Samuel, 'Co-motion: making space to Care for Country', *Geoforum*, 54 (2014), 132–41.

Mullins, Greg, *Firestorm* (Australia: Viking, 2021).

Musto, Josh, 'Earth is becoming "increasingly uninhabitable," scientists warn'. *The Independent*, 29 October 2024. Available at; <https://www.independent.co.uk/climate-change/news/earth-temperature-climate-uninhabitable-science-b2637796.html> (accessed 18 March 2025).

Nacu-Schmidt, Ami, Pearman, Olivia, Katzung, Jennifer and Boykoff, Max, 'This historic decline in emissions is happening for all the wrong reasons…', *MECCO Monthly Summaries*, Issue 40 (May 2020), <https://scholar.colorado.edu/concern/articles/wm117p983>, accessed 6 April 2025.

Nalau, Johanna, Blair Preston and Melissa Maloney, 'Is adaptation a local responsibility?', Environmental science & policy, 48 (2015), 89–98, accessed 4 October 2025.

Nalau, Johanna and Howden, Mark, 'Climate change is causing ever more disruption: can Australia's new adaptation plan help?', *The Conversation*, 15 September 2025, available at <https://theconversation.com/climate-change-is-causing-ever-more-disruption-can-australias-new-adaptation-plan-help-265276>, accessed 4 October 2025.

Narain, Sunita, *Conflicts of interest: my journey through India's green movement* (Penguin India, 2019), <https://www.penguin.co.in/book/conflicts-of-interest/>, accessed 4 October 2025.

Naudé, Wim, *Economic growth and societal collapse: beyond green growth and degrowth fairy tales* (Switzerland: Palgrave Macmillan, 2023).

NASA Science (NASA), 'The effects of climate change', [no date], available at <https://science.nasa.gov/climate-change/effects/> (accessed 18 March 2025).

Neff, Kristin D., Self-compassion: the proven power of being kind to yourself (New York: William Morrow, 2011).

Newell, Peter, and Dustin Mulvaney, 'The political economy of the just transition', *The geographical journal*, 179/2 (2013), 132–40.

Nixon, Rob, *Slow Violence and the Environmentalism of the Poor* (Cambridge, MA: Harvard University Press, 2011).

Noble, Bill, 'Should Australia ban fossil fuel advertising?', *The Australia Institute* (7 July 2024), <https://australiainstitute.org.au/post/should-australia-ban-fossil-fuel-advertising/>, accessed 25 October 2025.

Norris, Fran H., Susan P. Stevens, Betty Pfefferbaum, Karen F. Wyche and Rose L. Pfefferbaum, 'Community resilience as a metaphor, theory, set of capacities, and strategy for disaster readiness', *American Journal of Community Psychology* 41/1 (2008), 127–50.

Noor, Dhriti, 'University funding from fossil fuels slowing switch to green energy – report', *The Guardian* (5 September 2024), https://www.theguardian.com/business/article/2024/sep/05/universities-fossil-fuel-funding-green-energy>, accessed 6 April 2025.

NSW Department of Planning and Environment (DPE), NARCliM climate change projections for New South Wales (2021), available at <https://www.climatechange.environment.nsw.gov.au/climate-projections> (accessed 17 March 2025).

O'Brien, Karen, 'Is the 1.5°C target possible? exploring the three spheres of transformation', *Current opinion in environmental sustainability*, 31 (2018), 153–60, <https://doi.org/10.1016/j.cosust.2018.04.010>, accessed 4 October 2025.

Opper, Ilona, Mandy Maher, Jane Figgis and Sue Clifford, *Research: social connections – Neighbourhood Connect* (Melbourne: Neighbourhood Connect, 2023). <https://www.neighbourhoodconnect.org.au/research>, accessed 23 October 2025.

Oreskes, Naomi, Why trust science? (Princeton: Princeton University Press, 2018).

Oreskes, Naomi, and Erik M. Conway, Merchants of doubt: how a handful of scientists obscured the truth on issues from tobacco smoke to climate change (New York: Bloomsbury Press, 2010).

Organisation for Economic Co-operation and Development (OECD), *Climate finance provided and mobilized by developed countries in 2013–2022: climate finance and the USD 100 billion goal* (Paris: OECD Publishing, 2024), <https://doi.org/10.1787/19150727-en>, accessed 10 November 2025.

Organisation for Economic Co-operation and Development (OECD), *Scaling finance and investment for climate adaptation* (Paris: OECD Publishing, 2024), <https://www.oecd.org/en/publications/scaling-finance-and-investment-for-climate-adaptation_eeec8b52-en.htm>, accessed 10 November 2025.

Organisation for Economic Co-operation and Development (OECD), *Towards green growth* (Paris: OECD, 2011).

Otte, Hannah, 'The lawsuits that could save the world: inside climate courtroom battles', *Climate Cosmos* (13 June 2025), <https://climatecosmos.com/climate-news/the-lawsuits-that-could-save-the-world-inside-climate-courtroom-battles/>, accessed 12 March 2025.

Owens, Mick, and Maryanne Fuller, *2022 NSW flood inquiry: final report* (Sydney: New South Wales Government, 2022).

Oxfam, Confronting carbon inequality: putting climate justice at the heart of the COVID-19 recovery (Oxford: Oxfam International, 2020), <https://policy-practice.oxfam.org/resources/confronting-carbon-inequality-putting-climate-justice-at-the-heart-of-the-covid-62>, accessed 8 August 2025.

Oxfam, Who is responsible for climate change? (3 December 2023), <https://www.oxfamamerica.org/explore/stories/who-is-responsible-for-climate-change/>, accessed 7 April 2025.

Oxfam International, Confronting carbon inequality: putting climate justice at the heart of the COVID-19 recovery (Oxfam International, 2020), <https://policy-practice.oxfam.org/resources/confronting-carbon-inequality-putting-climate-justice-at-the-heart-of-the-covid-62>, accessed 12 May 2025.

Pahl, Sabine, Sheppard, Stephen, Boomsma, Christine and Groves, Chris, 'Perceptions of time in relation to climate change', *WIREs Climate Change* 5 (2014), 375–388.

Park, Crystal L., 'Making sense of the meaning literature: an integrative review of meaning making and its effects on adjustment to stressful life events', *Psychological Bulletin* 136/2 (2010), 257–301.

Pascht, Ariadne, Pierre Gubry and Cecilia Anckar, 'Marine conservation in Vanuatu: local conceptualisation and "assemblage"', *Ambio*, 51 (2022), 2389–2400, <https://doi.org/10.1007/s13280-022-01767-3>, accessed 4 October 2025.

Pascoe, Michael C., David R. Thompson, Zachary M. Jenkins and Chantal F. Ski, 'Mindfulness mediates the physiological markers of stress: systematic review and meta-analysis', *Journal of Psychiatric Research* 95 (2017), 156–78.

Patel, Shahrzad S., Michael B. Rogers, Richard Amlôt and G. James Rubin, 'What do we mean by "community resilience"? a systematic literature review', *PLOS Currents Disasters* (2017), <https://doi.org/10.1371/currents.dis.31b5bced32108faab5ee0ff7f4f5799e>, accessed 4 October 2025.

Pelizzon, Alessandro, *Ecological jurisprudence: the law of nature and the nature of law* (Singapore: Springer Singapore, 2025).

Pelling, Mark, Adaptation to climate change: from resilience to transformation (Routledge, 2011), <https://doi.org/10.4324/9780203889046>, accessed 4 October 2025.

Pelling, Mark, and Dill, Kathleen, 'Disaster politics: tipping points for change in the adaptation of sociopolitical regimes', Progress in Human Geography 34/1 (2010), 21–37.

Pelling, Mark, Karen O'Brien and David Matyas, 'Adaptation and transformation', *Climatic change*, 133 (2015), 113–27, <https://doi.org/10.1007/s10584-014-1303-0>, accessed 4 October 2025.

Pihkala, Panu, 'The process of eco-anxiety and ecological grief: a narrative review and a new proposal', *Sustainability*, 14/24 (2022), article 16628, <https://doi.org/10.3390/su142416628>, accessed 4 October 2025.

Pinker, Steven, Enlightenment now: the case for reason, science, humanism, and progress (New York: Viking Press, 2018).

Plan C, 'About the Northern Rivers Community Resilience Alliance' (no date), <https://www.planc.org.au/nr-community-resilience-alliance>, accessed 8 August 2025.

Plan C, *Community carers and responders* webpage (no author, no date), <https://www.planc.org.au/community-carers-and-responders>, accessed 4 October 2025.

Plan C, 'Facing Up: how do we live in the midst of the climate and ecological crises?', *Plan C* (no date), <https://www.planc.org.au/facingup>, accessed 26 October 2025.

Pester, Patrick, 'When did scientists first warn humanity about climate change?', *Live Science* (12 December 2021), <https://www.livescience.com/humans-first-warned-about-climate-change>, accessed 12 March 2025.

Piketty, Thomas, *Capital in the twenty-first century* (Cambridge, MA: Harvard University Press, 2014).

Powell, John A., Stephen Menendian and Wendy Ake, *Targeted universalism: policy & practice* (Berkeley: Haas Institute for a Fair and Inclusive Society, University of California, Berkeley, 2019).

Price, David, 'The sorrows of our time' (21 March 2025), *Medium*, <https://davidprice-26453.medium.com/the-sorrows-of-our-time-e74ec624425b>, accessed 4 October 2025.

Productivity Commission, Natural disaster funding arrangements, inquiry report no. 74 (Canberra: Productivity Commission, 2014).

Putnam, Robert D., Bowling alone: the collapse and revival of American community (New York: Simon & Schuster, 2000).

Quarantelli, Enrico, L., 'Emergent citizen groups in disaster preparedness and recovery activities' (1984), <http://udspace.udel.edu/handle/19716/1206>, accessed 4 October 2025.

Raworth, Kate, *Doughnut economics: seven ways to think like a 21st-century economist* (London: Random House Business, 2017).

Readfearn, Graham. 'Australia's biggest gas project greenlit to 2070 with "partial" protection for Indigenous rock art'. *The Guardian*, 12 September 2025. Available at: <https://www.theguardian.com/environment/2025/sep/12/north-west-shelf-gas-project-extended-to-2070-with-partial-protection-for-indigenous-rock-art> (accessed 18 March 2025).

Readfearn, Graham, and Adam Morton, '"climate villain": scientists say Rupert Murdoch wielded his media empire to sow confusion and doubt', *The Guardian* (23 September 2023). <https://www.theguardian.com/media/2023/sep/23/rupert-murdoch-climate-change-denial>, accessed 25 April 2025.

Renouf, Jean S. 'Making sense of climate change: the lived experience of experts'. *Climatic change* 164 (2021). Available at: <https://doi.org/10.1007/s10584-021-02986-5> (accessed 18 March 2025).

Resilient Kurilpa website (no author, no date), <https://www.resilientkurilpa.com/>, accessed 8 August 2025.

Richardson, Katherine, et al. 'Earth beyond six of nine planetary boundaries'. *Science advances* 9/37 (2023), eadh2458. Available at: <https://www.science.org/doi/10.1126/sciadv.adh2458> (accessed 18 March 2025).

Rigby, Campbell W., Allan Rosen, Helen L. Berry and Clive R. Hart, 'If the land's sick, we're sick: the impact of prolonged drought on the social and emotional well-being of Aboriginal communities in rural New South Wales', *Australian Journal of Rural Health* 19/5 (2011), 249–54.

Rimon, Akka, and Anote Tong, 'The seas are coming for us in Kiribati. will Australia rehome us?' (23 November 2021), *The Conversation*, <https://theconversation.com/the-seas-are-coming-for-us-in-kiribati-will-australia-rehome-us-172137/>, accessed 4 October 2025.

Riofrancos, Thea, 'Seize and resist: the global supply chain is up for grabs', *The baffler*, no. 54 (November 2020), <https://thebaffler.com/salvos/seize-and-resist-riofrancos>, accessed 4 October 2025.

Ripple, William J., Christopher Wolf, Timothy M. Lenton, Jay W. Gregg, Susan M. Natali, Philip B. Duffy, Johan Rockström and Hans Joachim Schellnhuber, 'Many risky feedback loops amplify the need for climate action'. Alliance of World Scientists, 2023. Available at: <https://scientistswarning.forestry.oregonstate.edu/climate-feedback-loops-project> (accessed 18 March 2025).

Rittel, Horst W. J., and Webber, Melvin M., 'Dilemmas in a general theory of planning', *Policy Sciences* 4/2 (1973), 155–169.

RMIT Innovation Catalyst and ANU Global Institute for Women's Leadership, *Climate change reporting in the Australian media: data for good insights paper* (November 2024, RMIT), <https://giwl.anu.edu.au/files/2024-11/ssrn-4851963.pdf>, accessed 6 April 2025.

Rose, Deborah Bird, *Nourishing terrains: Australian Aboriginal views of landscape and wilderness* (Canberra: Australian Heritage Commission, 1996).

Roy, Arundhati, *An ordinary person's guide to empire* (Cambridge, MA: South End Press, 2004), 86.

Royal Commission into National Natural Disaster Arrangements, *report* (Canberra: Australian government, 2020), <https://www.royalcommission.gov.au/system/files/2020-12/RoyalCommissionintNationalNaturalDisasterArrangements-Reportacccssible.pdf>

Saltmarsh, Chris, 'Deep adaptation – or climate justice?' (1 February 2022), *The Ecologist*, <https://theecologist.org/2022/feb/01/deep-adaptation-or-climate-justice>, accessed 20 August 2025.

ScienceDaily, 'Earth has hit its first climate tipping point, scientists warn'. 29 October 2025. Available at: <https://www.sciencedaily.com/releases/2025/10/251029002920.htm> (accessed 30 November 2025).

Schmelzer, Matthias, Vetter Andrea, and Vansintjan Aaron, *The future is degrowth: a guide to a world beyond capitalism* (London and New York: Verso Books, 2022).

School Strike 4 Climate, *School Strike 4 Climate website* (no date), <https://www.schoolstrike4climate.com/>, accessed 12 March 2025.

Seddon, Nathalie, Anne Chausson, Pam Berry, Cecile A. Girardin, Alison Smith and Ben Turner, 'Understanding the value and limits of nature-based solutions to climate change and other global challenges', *Philosophical transactions of the Royal Society B*, 375/1794 (2020), article 20190120, <https://doi.org/10.1098/rstb.2019.0120>, accessed 8 August 2025.

Serdeczny, Olivia, Sophia Adams, Flavio Baarsch et al., 'Climate change impacts in sub-Saharan Africa: from physical changes to their social repercussions', *Regional environmental change*, 17 (2017), 1585–1600, <https://doi.org/10.1007/s10113-015-0910-2>, accessed 8 August 2025.

Servigne, Pablo, and Stevens, Raphaël, *How Everything Can Collapse: A Manual for Our Times* (London: John Wiley & Sons, 2020).

Servigne, Pablo, Stevens, Raphaël and Chapelle, Gauthier, Another end of the world is possible: living the collapse (and not merely surviving it) (Cambridge: Polity Press, 2021).

Sharma, Aisha, 'Australia's criminalisation of peaceful protest is a stain on our democracy', *Crikey* (24 July 2024), <https://www.crikey.com.au/2024/07/24/climate-protest-criminalised-australia/>, accessed 12 March 2025.

Shiva, Vandana, Earth democracy: justice, sustainability, and peace (Cambridge, MA: South End Press, 2005).

Shute, Nancy. 'When attacks on science threaten our survival'. *Science News*, 2 May 2021. <https://www.sciencenews.org/article/when-attacks-on-science-threaten-our-survival>.

Siders, A. R., 'Managed retreat in the United States', *One earth*, 1/2 (2019), 216–25, <https://doi.org/10.1016/j.oneear.2019.09.008>, accessed 4 October 2025.

Simpson, Leanne Betasamosake, *As we have always done: indigenous freedom through radical resistance* (Minneapolis: University of Minnesota Press, 2017).

Smith, Linda Tuhiwai, Decolonizing methodologies: research and Indigenous peoples, 2nd edn (London: Zed Books, 2012).

Singh, Harjeet. '"They're expanding fossil fuels": Indian activist Harjeet Singh on rich countries'. Interview for *Democracy Now!*, 21 November 2024. Available at: <https://www.democracynow.org/2024/11/21/cop_29_finance_agreement_harjeet_singh> (accessed 18 March 2025).

Solnit, Rebecca, *Paradise built in hell: the extraordinary communities that arise in disaster* (New York: Viking Press, 2009).

Sovacool, Benjamin K., 'Hard and soft paths for climate change adaptation', *Climatic change*, 104/3–4 (2011), 381–413, <https://doi.org/10.1007/s10584-010-9984-5>, accessed 8 August 2025.

Spade, Dean, Mutual aid: building solidarity during this crisis (and the next) (London: Verso, 2020).

Sparrow, Jeff, Crimes against nature: capitalism and global heating (Melbourne: Scribe Publications, 2021).

Spira, Rupert, *Being aware of being aware* (Oxford: Sahaja Publications, 2017).

Statista, 'Atmospheric co$_2$ ppm by year 1959–2024'. 7 August 2025. Available at: <https://www.statista.com/statistics/1091926/atmospheric-concentration-of-co2-historic/> (accessed 18 October 2025).

Steffen, Will, Rockström, Johan, Richardson, Katherine, Lenton, Timothy M., Folke, Carl, Liverman, Diana, Summerhayes, Colin P., Barnosky, Anthony D., Cornell, Sarah E., Crucifix, Michel, Donges, Jonathan F., Fetzer, Ingo, Lade, Steven J., Scheffer, Marten, Winkelmann, Ricarda and Schellnhuber, Hans Joachim, 'Trajectories of the Earth System in the Anthropocene', *Proceedings of the National Academy of Sciences of the United States of America* 115/33 (2018), 8252–8259.

Stockholm Resilience Centre. *Planetary boundaries*. 2024. Available at: <https://www.stockholmresilience.org/research/planetary-boundaries.html> (accessed 18 March 2025).

Story Preservation Initiative, 'Terry Tempest Williams, Seeing beyond our time' (25 February 2015), <https://storypreservation.wordpress.com/2015/02/25/terry-tempest-williams-seeing-beyond-our-time/>, accessed 4 October 2025.

Ström, Tomas E., 'Digital disenchantment: Postcapitalism, AI and the end of the world', *Arena Quarterly* 5 (2021), 35–42.

Suls, Jerry, Paula Green and Sadia Hillis, 'Emotional reactivity to everyday problems, affective inertia, and neuroticism', *Personality and Social Psychology Bulletin* 29/7 (2013), 898–908.

Sultana, Farhana, 'The unbearable heaviness of climate coloniality', *Political geography*, 99 (2022), article 102638, <https://doi.org/10.1016/j.polgeo.2022.102638>, accessed 4 October 2025.

Sun, Evie, 'The impact of technology on our epidemic of loneliness'. [No date]. Available at: <https://brainmindsociety.org/posts/the-impact-of-technology-on-our-epidemic-of-loneliness> (accessed 18 March 2025).

Sütterlin, Nicole A., 'History of trauma theory', in *The Routledge Companion to Literature and Trauma*, ed. by Colin Davis and Hanna Meretoja (Abingdon: Routledge, 2020), pp. 11–22. Available at: <https://www.taylorfrancis.com/chapters/edit/10.4324/9781351025225-1/history-trauma-theory-nicole-s%C3%BCtterlin> (accessed 18 March 2025).

Swyngedouw, Erik, 'Apocalypse forever? Post-political populism and the spectre of climate change', *Theory, Culture & Society* 27/2–3 (2010), 213–32.

Tedeschi, Richard G., and Lawrence G. Calhoun, ‘Posttraumatic growth: conceptual foundations and empirical evidence’, *Psychological Inquiry* 15/1 (2004), 1–18.

Termeer, Catrien J. A. M., Dewulf, Art, and Breeman, Gerard E., ‘Governance of wicked climate adaptation problems’, in Knieling, Jürgen and Leal Filho, Walter L. (eds), *Climate Change Governance* (Berlin: Springer, 2013), pp 27–39.

The Australia Institute, ‘All-time high of 75 % of Australians concerned about climate change’, Media Release (13 October 2021), <https://australiainstitute.org.au/post/all-time-high-of-75-of-australians-concerned-about-climate-change/\>, accessed 27 October 2025.

The Australia Institute, ‘Climate crisis escalates cost-of-living pressures’, media release, 17 February 2025, available at <https://australiainstitute.org.au/post/climate-crisis-escalates-cost-of-living-pressures/> (accessed 18 March 2025).

The Australia Institute, ‘Local governments face soaring cost of climate change’, 10 September 2025, available at <https://australiainstitute.org.au/post/local-governments-face-soaring-cost-of-climate-change/> (accessed 28 September 2025).

The Australia Institute. ‘Open letter calls on government to set timeline for end of new fossil fuel projects’. 25 September 2024. Available at: <https://australiainstitute.org.au/post/open-letter-calls-on-government-to-set-timeline-for-end-of-new-fossil-fuel-projects/> (accessed 18 March 2025).

The New Daily, ‘it’s time to address how we actually survive climate change’, *The New Daily* (9 November 2025), <https://www.thenewdaily.com.au/opinion/2025/11/09/cop30-australia-climate>, accessed 13 November 2025.

The Returning, ‘Our story’ (n.d.). Available at: <https://www.thereturning.com.au/our-story>, accessed 23 October 2025.

Thich Nhat Hanh, *Peace is every step: the path of mindfulness in everyday life* (London: Random House, 2010).

Tillich, Paul, The courage to be (New Haven and London: Yale University Press, 2014) (original work published 1952).

Timperley, Jocelyn, ‘Who is really to blame for climate change?’, *BBC Future* (19 June 2020), <https://www.bbc.com/future/article/20200618-climate-change-who-is-to-blame-and-why-does-it-matter>, accessed 6 April 2025.

Tolle, Eckhart, The power of now: a guide to spiritual enlightenment (London: Yellow Kite, 2016).

Tsing, Anna Lowenhaupt, *The mushroom at the end of the world: on the possibility of life in capitalist ruins*. Princeton, NJ: Princeton University Press, 2015.

Thunberg, Greta, ‘Address at World Economic Forum: “Our house is on fire” – Jan 25, 2019’ (Archives of Women’s Political Communication, 25 January 2019), <https://awpc.cattcenter.iastate.edu/2019/12/02/address-at-davos-our-house-is-on-fire-jan-25-2019>, accessed 26 October 2025.

United Nations, ‘Five ways the climate crisis impacts human security’, *United Nations Climate Action* (2025), <https://www.un.org/en/climatechange/science/climate-issues/human-security>, accessed 10 October 2025.

United Nations Environment Programme (UNEP), Adaptation gap report 2024: come hell and high water – as fires & floods hit the poor hardest, it is time for the world to step up adaptation actions (Nairobi: UNEP, 2024), <https://www.unep.org/resources/adaptation-gap-report-2024>, accessed 10 November 2025.

United Nations Framework Convention on Climate Change (UNFCCC), *The Paris agreement* (2015; FCCC/CP/2015/10/Add.1)

United Nations Framework Convention on Climate Change (UNFCCC), *Report of the Conference of the Parties on its twenty-seventh session: establishment of the loss and damage fund* (Bonn: United Nations Framework Convention on Climate Change, 2023).

United Nations High Commissioner for Refugees (UNHCR), No escape: on the frontlines of climate change, conflict and forced displacement (Geneva: UNHCR, 2024).

United Nations Office for Disaster Risk Reduction (UNDRR), GAR 2025 hazard explorations: floods. global assessment report on disaster risk reduction (2025), available at <https://www.undrr.org/gar/gar2025/hazard-exploration/floods> (accessed 18 March 2025).

United Nations Office for the Coordination of Humanitarian Affairs, Libya: Flood update Flash Update No. 2 (15 September 2023) (New York: UNOCHA, 2023) <https://www.unocha.org/publications/report/libya/libya-flood-update-flash-update-no2-15-september-2023>, accessed 25 March 2025.

United Nations Regional Information Centre for Western Europe (UNRIC), ‘Global warming: 90% of emissions heat absorbed by the ocean’, 25 September 2023, available at <https://unric.org/en/global-warming-90-of-emissions-heat-absorbed-by-the-ocean/> (accessed 18 March 2025).

University Centre for Rural Health, ‘Northern Rivers Community Resilience Alliance: a movement for resilience’ (2 September 2025), <https://ucrh.edu.au/northern-rivers-community-resilience-alliance-a-movement-for-resilience/>, accessed 8 August 2025.

University of Melbourne, ‘HILDA survey shows inequality rises to a high’, *Newsroom*, 2025, available at <https://www.unimelb.edu.au/newsroom/news/2025/march/hilda-shows-inequality-rises-to-a-high> (accessed 8 April 2025).

University of Melbourne, ‘Scientists warn heatwaves will intensify for 1,000 years even after net zero’. *Science Daily*, 20 November 2025. Available at: <https://www.sciencedaily.com/releases/2025/11/251120002600.htm> (accessed 18 March 2025).

van Eck, Christel W., and Toni G. L. A. van der Meer, 'Narratives of hope and concern? examining the impact of climate scientists' communication on credibility and engagement', *Public understanding of science*, 34/6 (2025), 734–51, <https://doi.org/10.1177/09636625251314159>, accessed 17 March 2025.

Vardy, Mike, 'The need for resilience' (20 November 2020), <https://medium.com/about-time/the-need-for-resilience-9060c914a9e1>, accessed 8 August 2025.

Wahl, Daniel C., *Designing regenerative cultures* (Axminster: Triarchy Press, 2016).

Walker, Brian, and David Salt, *Resilience practice: building capacity to absorb disturbance and maintain function* (Washington, DC: Island Press, 2012).

Wallace-Wells, David, *The uninhabitable earth: life after warming* (London: Penguin Books, 2019).

Walton, Kate, 'how "murdochracy" controls the climate debate in Australia', *Al Jazeera* (24 January 2020), <https://www.aljazeera.com/news/2020/1/24/how-murdochracy-controls-the-climate-debate-in-australia>, accessed 25 April 2025.

Waters, Elissa, Sophie Webber, Svenja Keele, Natalie Osborne, Lauren Rickards and Tayanah O'Donnell, 'Reimagining climate change research and policy from the Australian adaptation impasse', Environmental science & policy, 142 (2023), 144–52, <https://doi.org/10.1016/j.envsci.2023.01.014>, accessed 4 October 2025.

We-Alli website (no author, no date), <https://www.wealli.com.au/>, accessed 8 August 2025.

Weir, Jessica K., Catherine Stacey and Kara Youngetob, *The benefits of traditional fire management practices for biodiversity* (Canberra: CSIRO Ecosystem Sciences, 2011).

Wellman, Barry, and Milena Gulia, 'Virtual communities as communities: net surfers don't ride alone', in Barry Wellman, ed., *Networks in the global village: life in contemporary communities* (Boulder: Westview Press, 1999), 331–66.

Wheatley, Margaret J., Turning to one another: simple conversations to restore hope to the future (San Francisco: Berrett-Koehler Publishers, 2002).

White, Lisa and Rennie, Ellie =, 'Have we learned the lessons from the 2022 Lismore flood disaster?', *ABC News* (27 February 2023), <https://www.abc.net.au/news/2023-02-27/have-lessons-been-learned-from-2022-lismore-flood-disaster/101863278>, accessed 5 April 2025.

Whitmarsh, Lorraine, and Capstick, Stuart, 'Perceptions of climate change', in Clayton, Susan and Manning, Christie (eds), *Psychology and Climate Change: Human Perceptions, Impacts, and Responses* (Oxford: Academic Press, 2018), 13–33.

Whyte, Kyle Powys, 'Indigenous climate change studies: indigenizing futures, decolonizing the Anthropocene', *English Language Notes* 55/1–2 (2017), 153–62.

Whyte, Kyle Powys, 'Too late for Indigenous climate justice? ecological and relational tipping points', *Wiley Interdisciplinary Reviews: Climate Change* 11/1 (2020), e603.

Wilkowski, Benjamin M., and Michael D. Robinson, 'The cognitive basis of trait anger and reactive aggression: an integrative analysis', *Personality and Social Psychology Review* 12/1 (2008), 3–21.

Wise, Russel M., Ioan Fazey, Mark S. Smith, Sally E. Park, Hallie C. Eakin, Ernst A. Van Garderen and Bruce Campbell, 'Reconceptualising adaptation to climate change as part of pathways of change and response', *Global environmental change*, 28 (2014), 325–36, <https://doi.org/10.1016/j.gloenvcha.2013.12.002>, accessed 4 October 2025.

Wisner, Ben, Piers Blaikie, Terry Cannon and Ian Davis, *At risk: natural hazards, people's vulnerability and disasters*, 2nd edn (London: Routledge, 2004).

Wray, Britt, *Generation dread: finding purpose in an age of climate crisis* (Toronto: Knopf Canada, 2022).

Women's Environmental Leadership Australia (WELA), Gender, climate and environmental justice in Australia: WELA report (2024), <https://wela.org.au/gender-climate-report/>, accessed 17 March 2025.

World Bank, *Inclusive green growth: the pathway to sustainable development* (Washington, DC: World Bank, 2012).

World Bank Group (WBG), 'What you need to know about food security and climate change', 17 October 2022, available at <https://www.worldbank.org/en/news/feature/2022/10/17/what-you-need-to-know-about-food-security-and-climate-change> (accessed 18 March 2025).

World Health Organization (WHO), *Climate change and health* (2023, WHO), <https://www.who.int/news-room/fact-sheets/detail/climate-change-and-health>, accessed 6 April 2025.

World Inequality Lab, Climate change and the global inequality of carbon emissions, 1990–2020 (WIL Working Paper 2021/23, Paris School of Economics, 2021), <https://wid.world/document/climate-change-and-the-global-inequality-of-carbon-emissions-1990-2020-world-inequality-lab-wp-2021-23>, accessed 12 May 2025.

World Meteorological Organization, *State of the global climate 2020* (Geneva: World Meteorological Organization, 2021), <https://wmo.int/media/news/climate-change-indicators-and-impacts-worsened-2020>, accessed 25 November 2025.

World Meteorological Organization (WMO), 'WMO confirms 2024 as warmest year on record at about 1.55°c above pre-industrial level', 10 January 2025, available at <https://wmo.int/news/media-centre/wmo-confirms-2024-warmest-year-record-about-155degc-above-pre-industrial-level> (accessed 18 March 2025).

World Wildlife Fund, 'UN report shows human activity has changed the planet forever', *World Wildlife Fund News Stories*, 29 August 2021, <https://www.worldwildlife.org/news/stories/un-report-shows-human-activity-has-changed-the-planet-forever>, accessed 12 February 2025.

Wynes, Seth, and Kimberly A. Nicholas, 'The climate mitigation gap: education and government recommendations miss the most effective individual actions', *Environmental research letters*, 12 (2017), article 074024.

Young, Kirsty. 'Gabor Maté: Seven things we learned when he spoke to Kirsty Young'. *BBC Young Again*. [No date]. Available at: <https://www.bbc.co.uk/programmes/articles/3TdSk6dlDmFwVYFzB34vXZZ/gabor-mate-seven-things-we-learned-when-he-spoke-to-kirsty-young> (accessed 18 March 2025).

Xu, Chi, Timothy A. Kohler, Timothy M. Lenton, Jens-Christian Svenning and Marten Scheffer, 'Future of the human climate niche'. *Proceedings of the National Academy of Sciences* 117/21 (2020), 11350–55. Available at: <https://www.pnas.org/doi/10.1073/pnas.1910114117> (accessed 18 March 2025).

Yunkaporta, Tyson, Sand talk: how Indigenous thinking can save the world (Melbourne: Text Publishing Company, 2019).

Yurchak, Alexei, *Everything was forever, until it was no more: the last Soviet generation* (Princeton: Princeton University Press, 2005).

Zel, Zoe, 'The future of renewable energy: election promises for 2025', Australia Unwrapped (2 April 2025), <https://www.australiaunwrapped.com/the-future-of-renewable-energy-election-promises-for-2025/>, accessed 5 April 2025.

Notes on the authors

DR RICHARD HIL is Adjunct Professor in the School of Health Sciences and Social Work at Griffith University, Gold Coast, and Adjunct Professor in the Faculty of Law, Business and Arts at Southern Cross University. An academic, activist, and author of numerous books, his recent publications include *Whackademia: An Insider's Account of the Troubled University*, *Selling Students Short: Why You Won't Get the University Education You Deserve*, *The Sacking of Fallujah: A People's History* (with Ross Caputi and Donna Mulhearn), and *Transforming Universities in the Midst of Global Crisis: A University for the Commons* (with Kristen Lyons and Fern Thompsett). He is also a regular contributor to magazines, newspapers, and journals. His community-based work spans the coordination of community education projects, and he serves as a proud member of the Board of Plan C.

DR JEAN RENOUF is the founder and Chair of Plan C ('Our plan is the community'), a registered charity dedicated to strengthening community resilience and regeneration across Australia. He is also the CEO of Safer Future, a for-purpose resilience consultancy focused on building crisis-ready capability across leadership, coordination and systems, a firefighter with Fire and Rescue NSW, and a former university lecturer. He received the Australian National Emergency Medal and two NSW Premier's Citations for his work during the 2019-2020 bushfires and the 2022 Northern Rivers flooding events.

Before this, he spent fifteen years as an international aid worker in conflict zones and disaster-affected regions, including Afghanistan, the Central African Republic, Congo, Haiti, Iraq, North Korea, and the Occupied Palestinian Territories. These experiences, alongside years of work with communities in Australia, have shaped his understanding of collapse, civic life, and the forms of solidarity that emerge in times of upheaval.

He has published academic and non-academic work on catastrophic disruptions and pathways for collective response and is a sought after public speaker. A long-term mindfulness practitioner, nature lover, and father of two young boys, he brings to this book an awareness for the world they will inherit and a deep commitment to the possibilities of regeneration, community, and grounded agency.

Index